Errata

M.A. Hulsen and J. van der Zanden: "Problems, Analysis and solutions for viscoelastic flow", in: *Pieter Wesseling* (Ed.), Research in Numerical Fluid Mechanics (Notes on Numerical Fluid Mechanics, Vol. 17).
 ISBN 3-528-08090-6

- Page 78, the line

$$c_1^2\phi_1 + c_2^2\phi_2 - (\phi_t + v_1\phi_1 + v_2\phi_2)^2 = 0 , \quad (19)$$

should read

$$c_1^2\phi_1^2 + c_2^2\phi_2^2 - (\phi_t + v_1\phi_1 + v_2\phi_2)^2 = 0 , \quad (19)$$

- Page 78, the line

$$c_1^2\phi_1 + c_2^2\phi_2 - (v_1\phi_1 + v_2\phi_2)^2 = 0 . \quad (21)$$

should read

$$c_1^2\phi_1^2 + c_2^2\phi_2^2 - (v_1\phi_1 + v_2\phi_2)^2 = 0 . \quad (21)$$

- Page 80, the line

$$\frac{\partial v}{\partial x} - \frac{\partial q}{\partial y} = 0, \quad (27)$$

should read

$$\frac{\partial r}{\partial x} - \frac{\partial q}{\partial y} = 0, \quad (27)$$

- Page 80, the first line of equation (30)

$$\frac{\partial p}{\partial x} - \eta_s \frac{\partial q}{\partial x} - \eta_s \frac{\partial v}{\partial y} - \frac{\partial t_{xx}}{\partial x} - \frac{\partial t_{xy}}{\partial y} = \dots ,$$

should read

$$\frac{\partial p}{\partial x} - \eta_s \frac{\partial q}{\partial x} - \eta_s \frac{\partial r}{\partial y} - \frac{\partial t_{xx}}{\partial x} - \frac{\partial t_{xy}}{\partial y} = \dots ,$$

- Page 82, the first line of equation (37)

$$-\int_C \underline{n} \cdot ([\rho\underline{v}\underline{v} + p\underline{\underline{1}} - \underline{\underline{t}}] + \eta_s[\frac{\partial \underline{v}}{\partial n}]) \cdot \underline{\psi} dC +$$

should read

$$\int_C (-\underline{n} \cdot [\rho\underline{v}\underline{v} + p\underline{\underline{1}} - \underline{\underline{t}}] + \eta_s[\frac{\partial \underline{v}}{\partial n}]) \cdot \underline{\psi}$$

- Page 83, the line

$$[\underline{\tau}] = \eta_s[\frac{\partial \underline{v}}{\partial n}], \qquad (\underline{n} \cdot [\frac{\partial \underline{v}}{\partial n}] = 0),$$

should read

$$[\underline{\tau}] = - \eta_s[\frac{\partial \underline{v}}{\partial n}], \qquad (\underline{n} \cdot [\frac{\partial \underline{v}}{\partial n}] = 0),$$

Pieter Wesseling (Ed.)

Research in Numerical Fluid Mechanics

Notes on Numerical Fluid Mechanics
Volume 17

Volume 1 Boundary Algorithms for Multidimensional Inviscid Hyperbolic Flows (Karl Förster, Ed.)

Volume 2 Proceedings of the Third GAMM-Conference on Numerical Methods in Fluid Mechanics (Ernst Heinrich Hirschel, Ed.) (out of print)

Volume 3 Numerical Methods for the Computation of Inviscid Transonic Flows with Shock Waves (Arthur Rizzi/Henri Viviand, Eds.)

Volume 4 Shear Flow in Surface-Oriented Coordinates (Ernst Heinrich Hirschel/ Wilhelm Kordulla)

Volume 5 Proceedings of the Fourth GAMM-Conference on Numerical Methods in Fluid Mechanics (Henri Viviand, Ed.) (out of print)

Volume 6 Numerical Methods in Laminar Flame Propagation (Norbert Peters/ Jürgen Warnatz, Eds.)

Volume 7 Proceedings of the Fifth GAMM-Conference on Numerical Methods in Fluid Mechanics (Maurizio Pandolfi/Renzo Piva, Eds.)

Volume 8 Vectorization of Computer Programs with Applications to Computational Fluid Dynamics (Wolfgang Gentzsch)

Volume 9 Analysis of Laminar Flow over a Backward Facing Step (Ken Morgan/ Jaques Periaux/François Thomasset, Eds.)

Volume 10 Efficient Solutions of Elliptic Systems (Wolfgang Hackbusch, Ed.)

Volume 11 Advances in Multi-Grid Methods (Dietrich Braess/Wolfgang Hackbusch/ Ulrich Trottenberg, Eds.)

Volume 12 The Efficient Use of Vector Computers with Emphasis on Computational Fluid Dynamics (Willi Schönauer/Wolfgang Gentzsch, Eds.)

Volume 13 Proceedings of the Sixth GAMM-Conference on Numerical Methods in Fluid Mechanics (Dietrich Rues/Wilhelm Kordulla, Eds.)

Volume 14 Finite Approximations in Fluid Mechanics (Ernst Heinrich Hirschel, Ed.)

Volume 15 Direct and Large Eddy Simulation of Turbulence (Ulrich Schumann/ Rainer Friedrich, Eds.)

Volume 16 Numerical Techniques in Continuum Mechanics (Wolfgang Hackbusch/ Kristian Witsch, Eds.)

Volume 17 Research in Numerical Fluid Mechanics (Pieter Wesseling, Ed.)

Pieter Wesseling (Ed.)

Research in Numerical Fluid Mechanics

Proceedings of the 25th Meeting of the Dutch Association for Numerical Fluid Mechanics

Friedr. Vieweg & Sohn Braunschweig / Wiesbaden

CIP-Kurztitelaufnahme der Deutschen Bibliothek

Research in numerical fluid mechanics/Pieter Wesseling (ed.). – Braunschweig; Wiesbaden: Vieweg, 1987.
(Notes on numerical fluid mechanics; Vol. 17)
(Proceedings of the ... meeting of the Dutch Association for Numerical Fluid Mechanics; 25)
ISBN 3-528-08090-6

NE: Wesseling, Pieter [Hrsg.]; Kontaktgroep Numerieke Stromingsleer: Proceedings of the ...; 1. GT

Manuscripts should have well over 100 pages. As they will be reproduced photomechanically they should be typed with utmost care on special stationary which will be supplied on request. In print, the size will be reduced linearly to approximately 75 %. Figures and diagrams should be lettered accordingly so as to produce letters not smaller than 2 mm in print. The same is valid for handwritten formulae. Manuscripts (in English) or proposals should be sent to the general editor Prof. Dr. E. H. Hirschel, Herzog-Heinrich-Weg 6, D-8011 Zorneding.

The addresses of the editors of the series are given on the last page.

Produced by W. Langelüddecke, Braunschweig
Printed in Germany

ISSN 0179-9614

ISBN 3-528-08090-6

PREFACE

The Dutch Association for Numerical Fluid Mechanics (Kontaktgroep Numerieke Stromingsleer, KNSL) was founded in The Netherlands in November 1974. Since then, the Association has organized meetings twice a year. The present volume contains the proceedings of the 25th meeting, held on October 20, 1986, at Delft University of Technology.

The purpose of the KNSL is to provide an opportunity for researchers in numerical fluid mechanics to meet regularly and to inform each other about their research in an informal atmosphere. Presentations preferably describe work in progress, and discussion of unsolved problems and unresolved difficulties is encouraged. The working language is Dutch. Nevertheless, science and technology are worldwide activities, and therefore it was decided to publish the proceedings of the 25th meeting in English.

The nine contributions to the 25th meeting were selected by profs. A.I. van de Vooren, C.B. Vreugdenhil and the editor. These works are far from covering completely all activity in this field in this country, but they are typical of what is going on. A wide range of subjects is discussed, including fundamental aspects of spectral methods, solution methods for the Euler equations and aeronautical applications, viscous ship hydrodynamics, shallow water equations, viscous flows with capillary and non-Newtonian effects, and turbulent heat transfer with industrial applications.

The 25th meeting of the KNSL was supported financially by ECN (Netherlands Energy Research Foundation), MARIN (Maritime Research Institute Netherlands), NLR (National Aerospace Laboratory), WL (Delft Hydraulics Laboratory), VEG Gasinstituut, Delft University of Technology and University of Twente. This support is gratefully acknowledged.

December 1986 P. Wesseling

CONTENTS

TRENDS IN CFD FOR AERONAUTICAL 3-D STEADY APPLICATIONS: THE DUTCH SITUATION

J.W. Boerstoel, A.E.P. Veldman, J. van der Vooren and A.J. van der Wees

National Aerospace Laboratory NLR, Informatics Division,

Anthony Fokkerweg 2, 1059 CM Amsterdam, The Netherlands

SUMMARY

Current and mid-term developments in computational 3-D steady aerodynamics software at NLR, focusing on the efficient aerodynamic design of the next generation of transport aircraft, are surveyed on a global level. Following a brief review of the various levels of sophistication in physical flow modelling, and their relation to the aerodynamic design process in general, the major aerodynamic problem areas that are at present accessible to computational aerodynamics are discussed. The coherence in computational methods development is subsequently explained by showing how the methods cover a growing part of the aircraft operating range. Subsequently, the approach taken towards the development of the most advanced methods, based on the Euler and Reynolds-averaged Navier-Stokes equations, is discussed. Here the accents are on proven technology, uniformity of approach, block-structured boundary conforming grids, flexibility, robustness, and adaptive local grid refinement for physical relevance. It is shown that the developments discussed presuppose access to the computing power offered by the present and upcoming generation of modern vectorcomputers. Finally, the informatics aspects are discussed. It is explained, that the steadily growing amount of computational aerodynamics software needs definite measures to keep things under control. The general technical concept, which is currently being developed at NLR to stay in control, is briefly surveyed. This involves the management of methods as well as data, and the interaction with the user. Computers/workstations are embedded in an efficient communication network.

INTRODUCTION

The purpose of the present paper is to explain the current and mid-term developments at NLR in steady aerodynamics computations. Since the Dutch aircraft industry focuses on transport aircraft, the contents of the paper is restricted to applications for this type of aircraft.

The development of computational aerodynamics software to day has four aspects that are of prime importance. The first aspect is the decision which aerodynamic problems are going to be solved. This involves the aerodynamic configuration, the flight conditions, and the purpose why the problem must be solved. These factors then lead to the choice of a level of physical flow modelling, dependent on the technology available.
The second aspects is the choice of a numerical approach, which eventually must lead to the construction of a numerical algorithm that solves the problem. Also here the choice must be carefully made, because computational aerodynamics methods must be robust, fast, and above all flexible (in the sense that their range of application can grow evolutionary). Wrong choices can easily frustrate the delicate balance that usually exists between these factors.

The result of the above discussions and choices is discussed in the chapter on "current and mid-term developments". Occasionally it is possible to acquire the necessary software from elsewhere (e.g. by software exchange), but in most cases in-house development is mandatory. Such developments require as a rule large investments over comparatively long periods of time (number of years). Therefore decision making must be thorough.
Subsequently, the approach taken towards computational aerodynamics methods based on the Euler and the Reynolds-averaged Navier-Stokes equations will be discussed in a special chapter. This subject has been chosen, because Euler and Reynolds-averaged Navier-Stokes based methods are most advanced, and their development towards applications of real engineering interest has in fact only just begun.

The third aspect is access to the computing power required. This aspect is discussed in some detail in the chapter on "computing power required".

The fourth aspect concerns the efficient development of the necessary software (i.e. fast, and at low cost), the implementation of the software in the computational aerodynamics infrastructure for efficient usage, and finally the efficiency of further evolutionary developments and maintenance. Under the title "informatics aspects", a final chapter discusses this problem area from a number of angles, and explains the general technical concept that is being developed at NLR to handle the management of methods as well as data, and the user interaction; computers/workstations are embedded in an efficient communication network.

CURRENT AND MID-TERM DEVELOPMENTS

Steady flow calculations can be based on various levels of sophistication of the physical flow model. Also, two basic types of computation are generally required, viz. direct and inverse computation. Before discussing the current and mid-term developments in more detail, these two aspects will be reviewed briefly. The choice for a certain level of physical flow modelling, or the type of computation, generally depends on the application (purpose, and physical relevance required) and the technology available (numerical techniques, computing power). In a design environment, computational speed very often prevails over completeness in the modelling of the physics.

In aircraft aerodynamics the most sophisticated level of physical flow modelling is the Navier-Stokes equations, where only the small subgrid scale turbulent eddies are covered by (isotropic) turbulence modelling (level V). These equations model on a continuum basis all relevant flow phenomena, including large turbulent eddies. These equations are essentially time dependent (because of turbulence), and at the present stage of technology still out of reach. One level lower (level IV) are the Reynolds-averaged Navier-Stokes equations (and subsets of these like the thin layer equations and the so-called parabolized Navier-Stokes equations). In these equations turbulence is modelled completely, partly on a theoretical and partly on an empirical basis, whence steady processes exist and the steady equations have meaning. Further simplifications of the flow model require, that the flow domain be decomposed in subdomains where the viscous effects are important (boundary layers, wakes), and subdomains where the viscous effects are negligible (inviscid outer flow). Naturally, the flow solutions in the viscous and inviscid subdomain must be coupled. Three levels of coupling are usually distinguished, viz. no coupling, weak coupling, and strong coupling. No coupling means, that the inviscid flow is computed without taking the boundary layers and the wakes into account; boundary layers/wakes are computed afterwards using inviscid pressure distributions. Weak coupling means, that the inviscid flow and the boundary layer/wakes are computed alternatively in an iterative fashion. Strong coupling means, that the inviscid flow and the boundary layers/wakes are computed simultaneously. Strong coupling is mandatory if the boundary layer separates (i.e. reverse flow occurs).

In inviscid subdomains, three further levels of simplification of the flow model are of interest. Level III is the so-called Euler equations. These equations constitute the most complete inviscid flow model. One level lower, level II, is potential theory. Here the flow is assumed isentropic and irrotational, in order to allow the introduction of a velocity potential. In this case, the five conservation laws of physics (mass, 3*momentum, energy), reduce to the law of mass conservation. The assumptions are correct for subsonic flow, and reasonably accurate for transonic flow as long as shockwaves are weak. The advantage is the reduction of the number of dependent variables from five to one. The lowest level of sophistication, level I, is the so-called Prandtl-Glauert equation, which is the fully-linearized small-disturbance version of potential theory. As the equation is linear, the representation of shock waves is no longer possible.

In viscous subdomains, various subsets of the Reynolds-averaged Navier-Stokes equations, boundary layer equations (laminar, turbulent), thin layer equations, parabolized equations, or even the full equations, are of interest.

In aerodynamic aircraft design, two basic types of computations are required. The first type is the direct computation, in which the aircraft geometry is given and the flow is required. From the flow data, the aerodynamic characteristics required are derived. At present, such computations are feasible on the levels I (Prandtl/Glauert), II (potential theory), and III (Euler) of inviscid flow modelling, and the level of boundary layer equations (often in integral form). However, level IV (Reynolds-averaged Navier-Stokes, and subsets) is coming rapidly within reach. The second type is the inverse computation, in which some characteristics of the flow are given and the aircraft geometry is required within certain constraints. Such inverse computations are of great value in wing design. Then a favourable pressure distribution for cruise conditions is prescribed by an experienced aerodynamic designer, and the geometry of the wing is sought under such constraints as given planform, and minimum allowed thickness. Such computations put a high demand on computational speed, and therefore are at present only feasible on the levels I (Prandtl/Glauert) and II (potential theory), and the level of boundary layer equations (often in integral form).

Current and mid-term developments aim at the efficient aerodynamic design of the next generation of transport aircraft.

Methods are required for subsonic and transonic cruise conditions, as well as for take-off and landing conditions. The revived interest in propeller propulsion (in view of its prospects for lower fuel consumption) then leads to the requirement that these methods must be applicable to not only jet-aircraft, but also to propeller-aircraft. Since the aerodynamic integration of the propulsion system is an important aspect in the aerodynamic design considerations, it is necessary to have methods that can handle complex aircraft configurations, including wing, body, tail, nacelles, pylons, winglets, propeller slipstreams, jet exhaust plumes, etc. (Fig. 1). In case of take-off and landing, the complexity is even greater, because control surfaces must be simulated as well, and vorticity shedding plays an important role (Fig. 2).

An important goal in the aerodynamic design of transport aircraft today is to improve upon their aerodynamic effectivity. This involves both take-off and landing, as well as cruise conditions. The opportunities to improve aircraft performance under cruise conditions are constrained by necessary characteristics at low speed during take-off and landing. If better devices can be developed to increase the lift during take-off and landing, it is possible to reduce the drag under cruise conditions.

For subsonic cruise conditions, as well as take-off and landing conditions, a higher-order accurate panel method is currently being developed on the basis of the Prandtl/Glauert equation (level I). Mid-term developments aim at extending the higher-order panel method to a field panel method on the basis of potential theory (level II), and at the incorporation of turbulent boundary layer and wake effects (mainly on the wing) in such a way that separation is allowed (strong interaction).

For transonic cruise conditions, finite-volume methodology based on potential theory (level II) is currently being developed for direct [1], as well as inverse, computations. In the latter case, emphasis is on wing design. Also in this case, mid-term developments aim at the incorporation of turbulent boundary layer and wake effects (mainly on the wing) in such a way that separation is allowed (strong interaction).

For installation effects of propellers and jet engines, finite-volume methodology based on the Euler equations (level III) is currently being developed [2]. Here the first goal is the interaction of a wing/nacelle with a propeller-slipstream (Fig. 3). Mid-term developments will involve jet engine inlet and exhaust flows (Fig. 4), requiring extension to the Reynolds-averaged Navier-Stokes equations (or subsets, level IV).

Regarding the improvement of aerodynamic effectivity, the first step is the development of devices to increase the lift during take-off and landing by studying two-dimensional airfoil/slats/flaps configurations (Fig. 5). Currently, a higher order field panel method (based on potential theory, level II, [3]) is being extended with a turbulent boundary layer in the strong-interaction sense. Mid-term developments will involve methodology on the basis of the Reynolds-averaged Navier-Stokes equations (or subsets, level IV).

The above discussed current and mid-term development of methods is the consequence of a pertinent policy to cover the aircraft operating range to the best possible extent in view of the technological possibilities.

Consider figure 6, where a qualitative picture is given of the range of applicability of methods, based on the various levels (I through IV) of physical flow modelling, in the different parts of the operating range of a subsonic transport aircraft. Most aircraft flying today have been designed using mainly Prandtl/Glauert methods (level I), and of course the windtunnel. At best, also some early potential methods (level II) were used incidentally. For the design of the next generation of transport aircraft it is necessary to have sufficiently powerful methods on level I and levels II, III. These are basically the methods that must handle complex aircraft configurations in take-off, cruise and landing, and must facilitate the aerodynamic integration of the propulsion system. However, since these flight conditions cannot be investigated in sufficient detail without taking into account boundary layer separation effects to some extent (e.g. on the wing), it is also necessary to extend these methods onto the strong interaction levels IV/I and IV/II,III for specific applications. Also, e.g. jet engine exhaust flows can only be handled adequately by the Reynolds-averaged Navier-Stokes equations (or subsets thereof), level IV. Similarly, the development of devices to increase the lift during take-off and landing requires the Reynolds-averaged Navier- Stokes equations to model the physics adequately (level IV).

APPROACH TO EULER AND REYNOLDS-AVERAGED NAVIER-STOKES METHODOLOGY

The timely development of the comparatively new and complicated methodology associated with the Euler and Reynolds-averaged Navier-Stokes equations, at a reasonable cost, requires careful consideration of all aspects involved. In this chapter, some ideas that exist today at NLR, and also some decisions that have already been taken, will be briefly discussed.

The approach taken is first of all to use, whenever possible, proven technology, in order to cut down the development time and cost. New basic research will therefore be carried out only if the aerodynamic goals set cannot be reached on the basis of the proven technology available.

A second important issue is the awareness of the fact, that the Euler equations, and e.g. the thin-layer Navier-Stokes equations, or the parabolized Navier-Stokes equations, are all subsets of the full Reynolds-averaged Navier-Stokes equations. This leads to the general strategy, that a flow solver for the full Reynolds-averaged Navier-Stokes equations must function properly for all subsets. Naturally, this strategy sets a requirement for the development of an Euler flow solver which generally precedes the development of flow solvers for the viscous subsets of the Reynolds-averaged Navier-Stokes equations. It should be realized that a high demand for computational efficiency can easily lead to developments which depart from the above strategy. Though such developments cannot always be avoided, there occurrence should be minimized.

Thirdly, all software development is preferably directed to three-dimensional flow, right from the beginning. This point of view is taken, because the successfull generalization of a two-dimensional approach requires three-dimensional considerations anyway, and experience has taught that two-dimensional pilot versions of the software as a rule do not provide sufficient insight in the informatics aspects of three-dimensional flow simulation.

Finally, an integrated uniform approach is taken towards the problem areas of gridgeneration, flow solving, and visualization. In particular, the areas gridgeneration and flow solving will be discussed below in more detail.

The first choice in gridgeneration is always between fully boundary conforming, not boundary confirming at all, or a mixed form of these two extremals. Mainly based on two arguments, here the choice is made for fully boundary conforming grids. The first, and most important, argument has to do with the fact that the most difficult part of doing Euler and Reynolds-averaged Navier-Stokes calculations is not the development of a stable method, but rather the achievement of a physically relevant solution. Then, of course, high accuracy of boundary-condition implementation is a prerequisite, and it is firmly believed that such high accuracy can only be obtained using fully boundary conforming grids. The second, and subsidiary, argument is that, only in a fully boundary conforming grid, control can be exercised over the coordinate directions in the vicinity of the boundary. Such control can be important in using algebraic turbulence models such as mixing length models and eddy viscosity models. But also the "thin-layer" and "parabolized" versions of the Reynolds-averaged Navier-Stokes equations require such control explicitly.

For generating the fully boundary conforming grids, the following set-up has been decided upon [2]. First, the physical space surrounding the aircraft (or some of its components) is made finite by placing boundary surfaces, assuming that outside these surfaces the flow is known. Subsequently, this finite physical domain is subdivided into blocks in such a way that each block is topologically equivalent to a cube in computational space. Each block (cube) has six faces, twelve edges, and eight vertices. Though block-packing is restricted to "face-to-face", they can still be assembled to computational domains of arbitrary topological complexity (Fig. 7). Once the block-subdivision is established, and the relationships between all faces, edges, and vertices are determined, the grid is set up by the subsequent generation of gridpoint distributions in each edge, face and block, using (transfinite) linear interpolation. This leads to a grid with hexahedral cells and grid lines that are continuous across the faces, edges and vertices of adjacent blocks (Fig. 8). The final step is to smooth the grid in each block using an elliptic method that acts on user-provided information affecting the cell-size distribution. The set-up chosen is sufficiently flexible to allow generalizations such as slope-continuous gridlines or discontinuous changes in gridpoint distributions across block faces.

The advantages of the above described approach to gridgeneration become clear if flow solver development is considered in conjunction with the vector/parallel processing capabilities of modern supercomputers. By construction, the data corresponding to each block are well-ordered, and this is favourable from the point of view of efficiently approximating and solving the flow equations. This also contributes to the vectorizability of the flow solver algorithm. But also the fact that there can be drawn on an extensive literature on finite-difference/volume technology should not be forgotten. Finally, the accuracy of the approximation of the flow equations benefits from the smoothness of the grid in each block. Further advantages of the block-structuring are its amenability to parallel processing, and the inherent possibility of using a different subset of the Reynolds-averaged Navier-Stokes equations in each block.

The above argument shows that a block-structured grid of hexahedral cells is a flexible approach towards the development of Euler and Reynolds-averaged Navier-Stokes based methodology for complex three-dimensional aerodynamic shapes. As such it is considered to be an alternative to the often advocated unstructured grids using tetrahedral cells. Note, however, that a hexahedral cell can be subdivided into either five or six tetrahedral cells, whence the gridgeneration approach described above can be used also to generate an unstructured grid with tetrahedral cells.

With respect to the development of flow solvers for the Euler and Reynolds-averaged Navier-Stokes equations, the afore mentioned general strategy, viz. that a flow solver for the full Reynolds-averaged Navier-Stokes equations must function properly for all subsets, leads to the following considerations.

At present there are strong indications that the solution of the steady Reynolds-averaged Navier-Stokes equations may be hampered by non-existence as well as by non-uniqueness problems. Non-existence of steady solutions for two-dimensional laminar flow involving separation has been demonstrated in an asymptotic framework [4], and in interacting boundary-layer theory [5]. In [4], also non-uniqueness was found. For three-dimensional flow, non-existence of a steady solution of the Navier-Stokes equations was observed in [6]. The above reasons support the viewpoint that the notion of a steady solution should be replaced by the notion of a limit solution of the unsteady Reynolds-averaged Navier-Stokes equations as time goes on. Hence, regarding the building of flow solvers, there is a strong preference to base these on (pseudo) time integration of the unsteady equations. A bycoming advantage is then, that such solvers can be generalized to time-accurate unsteady flow (e.g. buffet) with comparative ease.

Another consideration of a general nature is a plea for adaptive local grid refinement. As observed before, the most difficult part of doing Euler and Reynolds-averaged Navier-Stokes calculations is the achievement of a physically relevant solution. Locally this will require an extremely fine grid to obtain the necessary accuracy. A good example in viscous flow is the resolution of shear layers of which the position is unknown beforehand. An example in inviscid flow (Euler equations) is the avoidance of spurious entropy production. However, it is mandatory that the total number of grid points be kept as low as possible from the point of view of acceptable computational time and cost. Two grid refinement strategies are possible. The first one is repositioning of a fixed number of grid-points; this strategy has the obvious disadvantage that a local increase in resolution is inevitably accompanied by a local decrease in resolution elsewhere, and can therefore easily lead to areas of too low resolution. The second one is the local insertion/deletion of gridpoints in an otherwise fixed grid. In light of the above discussion, the second strategy is definitely favoured.

More specific considerations with respect to flow solver development concern discretization and solution strategy.

Consider the subset Euler equations. These inviscid equations allow the occurrence of true discontinuities, viz. shockwaves and contact discontinuities. It is well known that the proper capture of such discontinuities by any numerical method requires discretization schemes which are fully conservative approximations of the Euler equations in full conservation form. With respect to the Reynolds-averaged Navier-Stokes equations, this requirement carries over to the convective parts of the equations. But even for flows without shockwaves or shear layers (contact discontinuities in inviscid flow), there is strong evidence that maintaining conservation in discretized form enhances accuracy considerably (such evidence comes from potential flow solutions of internal as well as external flow, [7]). In maintaining conservation in discretized form, grid discontinuities across block-faces, and similar discontinuities caused by adaptive local grid refinement, require special attention (Fig. 9).

A few considerations with respect to the solution strategy are the following. Time-explicit integration schemes are computationally simple and well amenable to vectorization. Time-implicit integration schemes are computationally definitely more complex and less amenable to vectorization. Also, with time-explicit integration schemes the allowable time-step is seriously limited by stability considerations (accuracy considerations are of interest only if a time-accurate solution is required). In general, the performance of a time integration scheme depends on the combined effect of its stability and vectorizability properties. In (almost) inviscid flow the allowable time-step of an explicit integration scheme is proportional to the local spatial mesh size; as time-accuracy is not required, the local allowable time-step can be used to accelerate convergence. Mainly based on their computational simplicity, and on their amenability to vectorization, time-explicit integration schemes are preferred over time-implicit integration schemes in this case. However, in viscous-dominated flows the situation changes, because in the viscous regions the local spatial mesh sizes have to be much smaller then in the inviscid region. This leads to a drastic reduction of the allowable time-step for explicit schemes. It is not believed that the good vectorization properties of time-explicit schemes are sufficient compensation for the indeed very small allowable time-steps. Hence, in this case there is a definite preference for time-implicit integration schemes per block. This way the scheme is still amenable to parallel processing. However, it should be realized, that the solution algorithm of a time-implicit integration scheme per block is seriously affected in case gridpoints are added locally for adaptive grid refinement.

Finally, the subject of convergence acceleration will be discussed in terms of cost effectiveness. Here the viewpoint will be taken, that a convergence acceleration technique for a given method is cost effective, provided that a solution of given accuracy is reached in significantly less computing time without having negative effects on the robustness and flexibility of the original method, and provided that the extra development time and cost are justified in the light of its usage. For the very complicated Euler and Reynolds-averaged Navier-Stokes flow solvers, that are required to treat the complex aerodynamic shapes associated with the design of future transport aircraft, this viewpoint is believed to limit the choice of convergence acceleration techniques to those computationally simpler than the original method, and equally vectorizable. Multigrid technique, which in practice has turned out to be complex, and to require high development cost, is certainly not among them. This does not preclude, however, the usage of multigrid technique for special well-defined applications.

Apart from the need for the computing power of a modern vectorcomputer (as illustrated in the next chapter) there is one common pacing item in the development of Euler and Reynolds-averaged Navier-Stokes methodology. This is building up the technology for adaptive local grid refinement, involving research in establishing the proper criteria for refinement, in devising a flexible discretization strategy, and in coping with the consequences for the solution process. Though not discussed in this paper, turbulence modelling is yet another important pacing item in the development of Reynolds-averaged Navier-Stokes methodology.

COMPUTING POWER REQUIRED

Integration of CFD methods as discussed above in the aerodynamic design process presupposes extensive testing and evaluation. Experience has taught that this process can only be carried out efficiently if full-scale calculations can be performed within, say, one half hour turn-around time. This requirement is based on the fact that the timely development (in a number of years) of such complicated methods demands that, on each working day, a number of full-scale computations can be executed and analyzed.

The above requirement is quantified in table 1 for two current developments and one mid-term development. The current developments are (1) the Euler finite-volume method to calculate propeller-slipstream/nacelle/wing interaction (propulsion system installation effects, see Fig. 3) and (2) the inverse potential finite-volume method to design the wing of a transport aircraft under transonic cruise conditions. The mid-term development is (3) the two-dimensional Reynolds-averaged Navier-Stokes method for airfoil/slats/flaps configurations (Fig. 5). Considering that future three-dimensional Reynolds-averaged Navier-Stokes calculations require a yet significantly larger computing power, this table shows that computing speed must be in the order of at least 300 Mflop/s, and the central memory must be larger than 40 M numbers.

Computing power such as indicated above is today only available in modern vectorcomputers such as e.g. CRAY-2, CRAY X-MP, NEC SX-2, and the upcoming ETA-10. Software development therefore presupposes short-term access to a supercomputer of this class.

INFORMATICS ASPECTS

The computational aerodynamics process involves repeated application of the following functions:

- Geometry definition (and manipulation), in order to obtain a mathematical representation of the geometry of the aerodynamic shapes. This function can be attributed to commercially available CAD/CAM packages.

- Grid generation (and manipulation), in order to cover the flow domain with a computational grid, accessible to a particular flow solver. As was already explained, flow solver developments are based on block-structured grids. The block subdivision of the flow domain, and the grid inside each block, are typical for the mathematical flow model used. This is dictated by the flow phenomena that the method must describe, and by the resolution required. Block subdivision is a process that can be carried out manually only in very simple cases. However, complex aerodynamic shapes can easily require in the order of one hundred blocks, and an automated process is mandatory. Commercially available CAD/CAM solid modelling packages might be of help. Since general purpose software for the subsequent generation of grids in the block structure, and for establishing their mutual relationships, are not (yet) commercially available, this is an in-house development.

- Flow calculation, in order to obtain a flow solution for a given aerodynamic shape, grid, mathematical model, and flow condition. Flow solvers are predominantly the subject of in-house developments. Commercial general purpose software is -generally speaking- not available for the many highly nonlinear problems of computational aerodynamics.

- Postprocessing for presentation and analysis of the computed flow. This involves both inspection (quick-look) postprocessing and analysis (detailed, complex-look) postprocessing. This function can only partly be accomplished by commercially available CAD/CAM and graphics packages, and therefore in-house developments are being carried out in parallel.

The above functions require different types of hardware equipment. Geometry definition (and manipulation), as well as postprocessing, are heavily interactive graphics applications, requiring at least workstation power. Grid generation requires both graphics and mainframe computing power, while flow calculation is a typical number crushing activity requiring a supercomputer.

In computational aerodynamics, a steadily graving amount of software is becoming available to perform the functions described above. In computational aerodynamic design processes, many different methods are being used, ranging from the simplest linear potential methods to highly complicated Reynolds-averaged Navier-Stokes methods; the choice of a flow solver depends on the balance between information required, its capabilities, and its computational cost. Growth can be observed both in latitude (more different methods and applications; more large scale applications, i.e. more gridpoints) as well as in depth (more complex physical modelling; increasing range of physical phenomena). As a result, a hierarchy of methods has developed, whereby each method has its own range of applicability (inviscid/viscous flow; rotational/irrotational flow; wing-alone/complex aircraft configuration) and its own computer resource requirements (little/much computing time; central processor/central memory usage). None of these methods can be used at a reasonable cost to cover all applications. Domain splitting concepts will lead to even more complicated methods, integrating e.g. full potential, Euler, boundary layer, and Reynolds-averaged Navier-Stokes methods. The amount of data absorbed and produced by present day computational aerodynamics methods, which is already tremendous, will grow even further with the new generation of vectorcomputers such as CRAY-2, CRAY X-MP, NEC SX-2, and the upcoming ETA-10.

Ways have to be found to stay in control of the above indicated developments. This involves the management of the methods, as well as of the modules of which they are composed. It also involves the management of the associated data. Finally, the interaction with the user and the embedding of computers/workstations in an efficient communication network are important aspects.

The general technical concept which is being developed at NLR to stay in control of e.g. the computational aerodynamics process, i.e. the hierarchy of the methods and modules of which the software is composed, the associated data, and the interaction with the user, is shown in figure 10. The concept requires that both methods, and modules, can be coupled on the functional level, while data-management is required to control the communication between the various methods. Method/module- and data-management can be automated when strict agreements are made on how the methods/modules in the hierarchy are to be interfaced, as well as strict rules are defined as to their individual usage.

For the management of methods, and of the modules of which they are composed, the system MEBAS (MEthod BAse System is being developed (in this framework a module is also called a method), see also [8]. MEBAS can operate (store, retrieve, couple) on a method-base of well-described methods, and is composed of two subsystems, viz. the method manager, and the executive. The method manager can be used for activities such as assemblage, repair, replacement, and versions management of methods, and of the modules of which these are composed. The executive takes care of the job execution task and operates on a library of executable methods (programs) built for specific applications.

Data-management is realized through the use of the system EDIPAS (Engineering Data Interactive Presentation and Analysis System), see [9]. Usage of EDIPAS requires that a common database be defined. The structure of this common database depends on the application to which an end-user applies the computational aerodynamics process. Each application can use its own database structure. As such, the common EDIPAS database forms the transfer point of data and the associated information between the various functions (geometry definition, grid generation, flow calculation, postprocessing) of the computational aerodynamics process (Fig. 11). All methods must therefore have a formal, application-dependent, interface with EDIPAS. It follows, that all methods can in principle communicate with each other via the common EDIPAS database, requiring only one interface for each method. An important advantage of the concept is also, that it requires of the software developer the careful a priori definition of the common database structure, of the EDIPAS interfaces, and of the control over each individual method.

EDIPAS can also partly perform the postprocessing function for presentation and analysis.

User interaction is realized using COLAS (COmmand Language System), see again [8].

The use of the above general technical concept (MEBAS, EDIPAS, COLAS) requires the proper definition of interfaces between methods, and as such avoids patchwork when integrating methods to applications. It also stimulates the reusability of software, and thus reduces software development costs.

CONCLUDING REMARKS

Current and mid-term developments in computational 3-D steady aerodynamics software at NLR have been shown to be directed towards the needs of the Dutch aircraft industry, and to cover a widening part of the transport aircraft operating range, crossing the separation onset boundary and occasionally protruding into the Reynolds-averaged Navier-Stokes range (Fig. 6). All levels of physical flow modelling, ranging from the linear Prantl-Glauert equation to the highly nonlinear Reynolds-averaged Navier-Stokes equations, are involved and have their specific area of application. There is in general a clear tendency towards complex geometries, which places a heavy accent on the integrated uniform approach to gridgeneration, flow solver development, and visualization. In this respect, a block-decomposition strategy has been accepted to generate boundary conforming structured grids with regular connectivity in each block. In the area of the well-established methods, based on (linearized) potential theory (with or without boundary layers), the emphasis is on complete aircraft in take-off, cruise, and landing configuration. In the area of the most advanced Euler and Reynolds-averaged Navier-Stokes based methods, the emphasis is first of all on installation effects of the propulsion system and high-lift devices to be used in take-off and landing. The development of Euler methods is well underway, while the development of Reynolds-averaged Navier-Stokes methods is about to start.

The approach towards Euler and Reynolds-averaged Navier-Stokes methods has been discussed. Time-dependent equations, adaptive local grid refinement, and fully-conservative finite volume schemes are favoured. Opinions are expressed with respect to the (pseudo) time integration of the scheme (explicit, implicit, multigrid). Pacing items are identified to be the various aspects of adaptive local grid refinement, and turbulence modelling.

The need for a modern vectorcomputer of the class CRAY-2, CRAY X-MP, NEC SX-2, ETA-10 is stressed.

A general technical informatics concept, which is being developed at NLR to remain in control of the rapidly expanding software and associated data in computational aerodynamics, is presented. This involves method-management, data-management, and user interaction. Computers/workstations are embedded in an efficient communication network.

REFERENCES

[1] Van der Vooren, J., Van der Wees, A.J., Meelker, J.H.: MATRICS, transonic potential flow calculations about transport aircraft, NLR MP 86019 U (1986).

[2] Boerstoel, J.W.: Problem and solution formulations for the generation of 3-D block-structured grids, NLR MP 86020 U (1986).

[3] Oskam, B.: Transonic panel method for the full potential equation applied to multi-component airfoils, AIAA paper 83-1855.

[4] Stewartson, K., Smith, F.T., Kaups, K.: Marginal separation, Stud. Appl. Math. 67 (1982), pp. 45-61.

[5] Henkes, R.A.W.M., Veldman, A.E.P.: Breakdown of steady and unsteady interacting boundary layers, NLR MP 86047 U (1986).

[6] Kordulla, W., Vollmers, H., Dallmann, U.: Simulation of three-dimensional transonic flow with separation past a hemisphere-cylinder configuration. Paper presented at the AGARD Symposium on Applications of Computational Fluid Dynamics in Aeronautics, 7-10 April 1986, Aix-en-Provence, France.

[7] Van der Vooren, J.: Private communication (1986).

[8] Loeve, W., Heerema, F.J., Van Hulzen, J.J.P.: The CAD environment of the National Aerospace Laboratory NLR, The Netherlands, NLR MP 86005 U (1986).

[9] Heerema, F.J., Van Hedel, H.: An engineering data management system for computer aided design, Adv. Eng. Software, Vol. 5, No. 2, pp. 67-75 (1983).

Table 1 Computing power required for current and mid-term method developmemt.

method	total flop	flop/s	CM	total DATA
1	135 - 360 G	80 - 200 M	10 - 40 M	3 - 25 M
2	70 - 280 G	40 - 160 M	5 - 30 M	1,5 - 6 M
3	200 - 500 G	110 - 280 M	1 M	1 M

1: propeller-slipstream/nacelle/wing interaction; Euler equations

2: transonic flow about a complex aircraft configuration under cruise conditions: inverse wing design; potential theory

3 airfoil/slats/flaps; two-dimensional Reynolds-averaged Navier-Stokes equations

G = giga = 10^9, M = mega = 10^6

total flop: total number of floating point operations required

flop/s : number of flop per second required for one half hour turn-around time

CM : central memory size required, expressed in 32 or 64 bit words

total DATA: size of dataset required for input preparation and output inspection, expressed in 32 or 64 bit words

Fig. 1 Examples of complex transport aircraft configurations

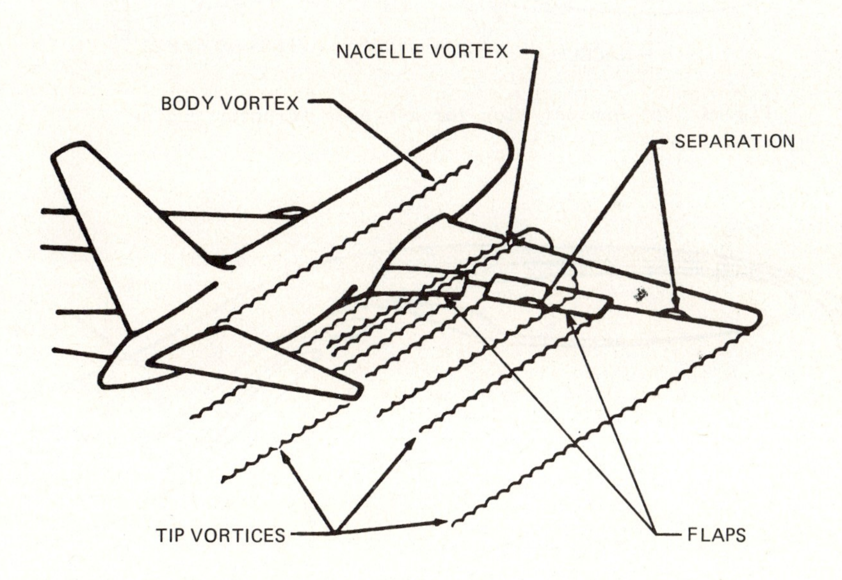

Fig. 2 Transport aircraft in take-off, or landing

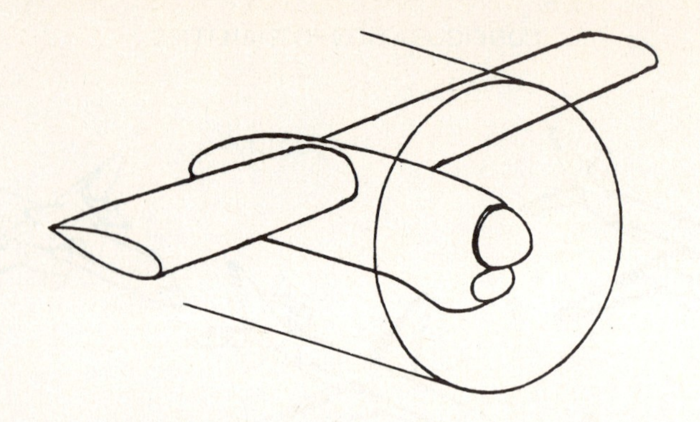

Fig. 3 Wing/nacelle/propeller-slipstream interaction

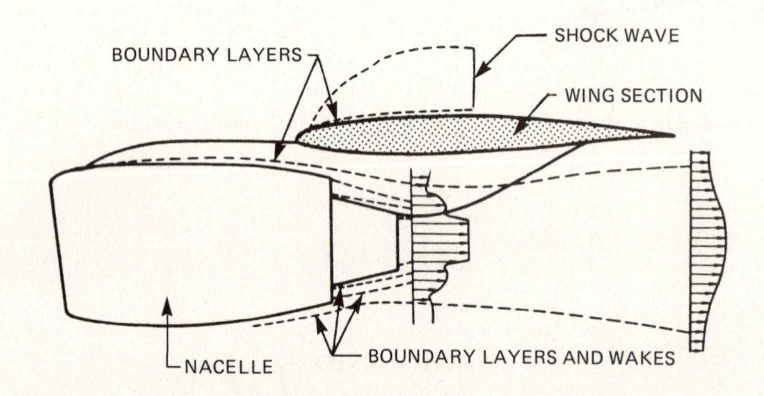

Fig. 4 Jet exhaust flow for a bypass turbofan engine

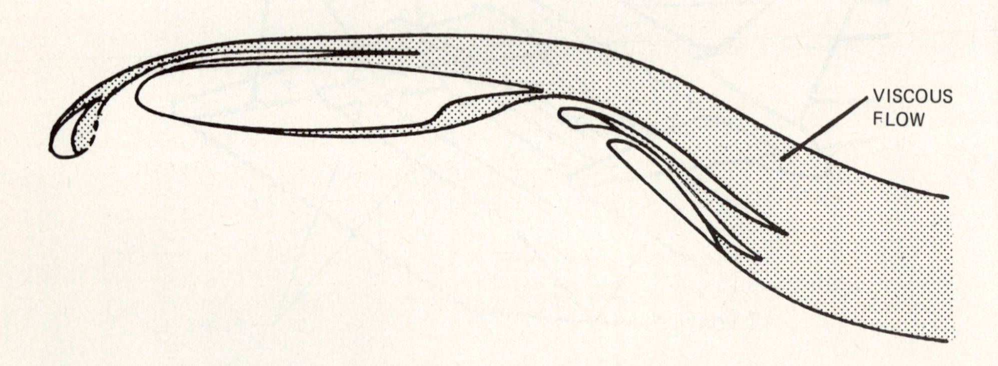

Fig. 5 Flow pattern for an airfoil/slat/flaps configuration

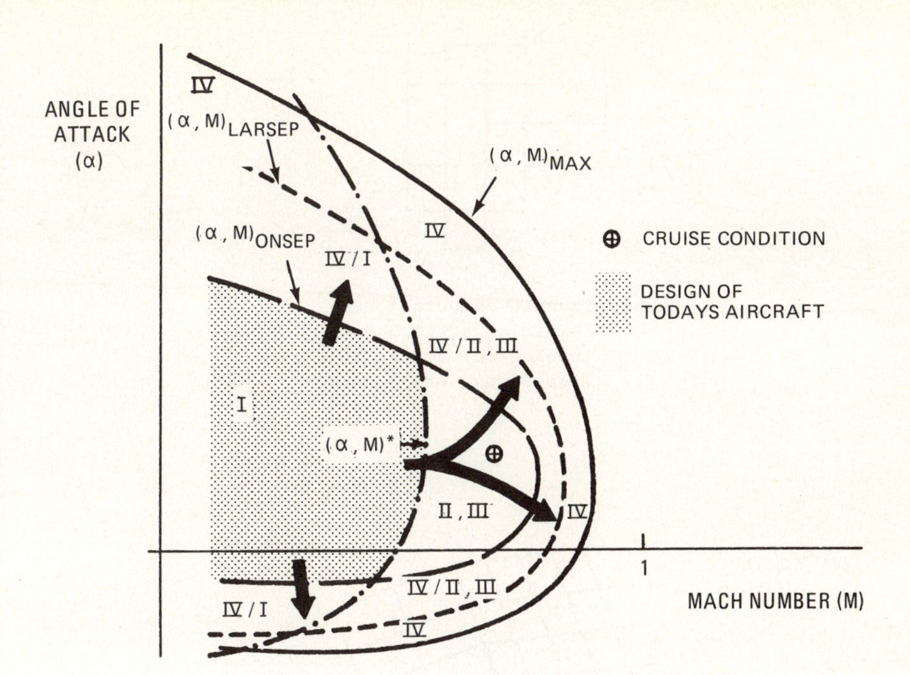

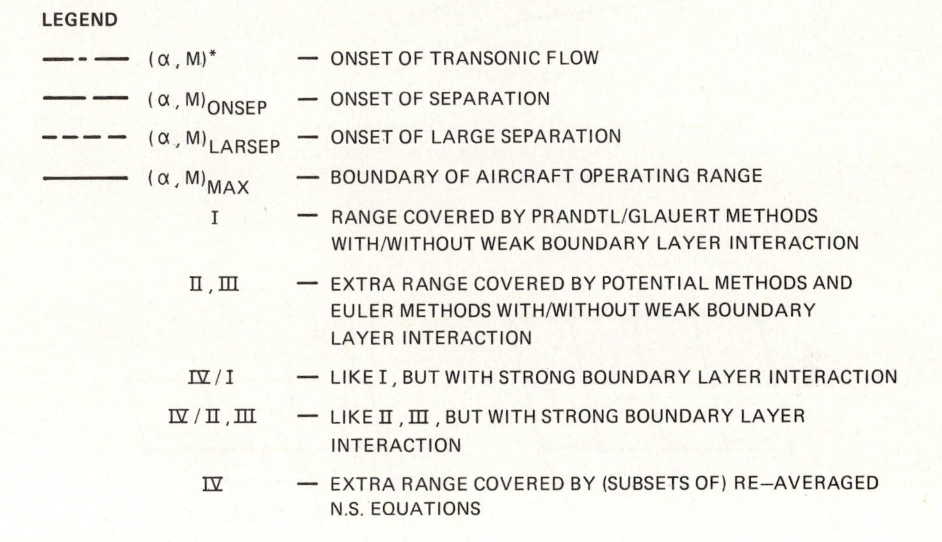

Fig. 6 Parts of the operating range of a subsonic transport aircraft covered by methods based on the various levels of physical flow modelling

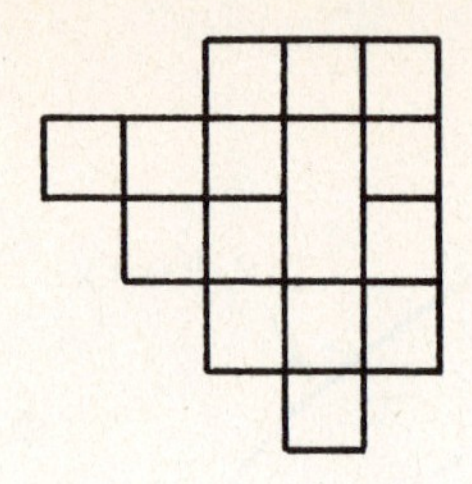

Fig. 7 Example computational domain of arbitrary topological conplexity (two-dimensional)

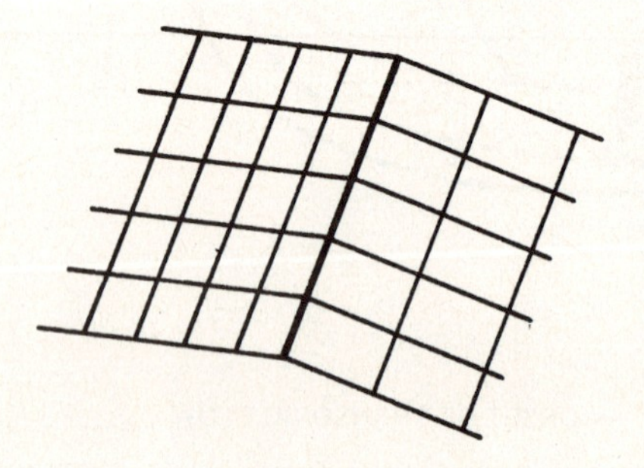

Fig. 8 Example of a continuous grid across block-faces (two-dimensional)

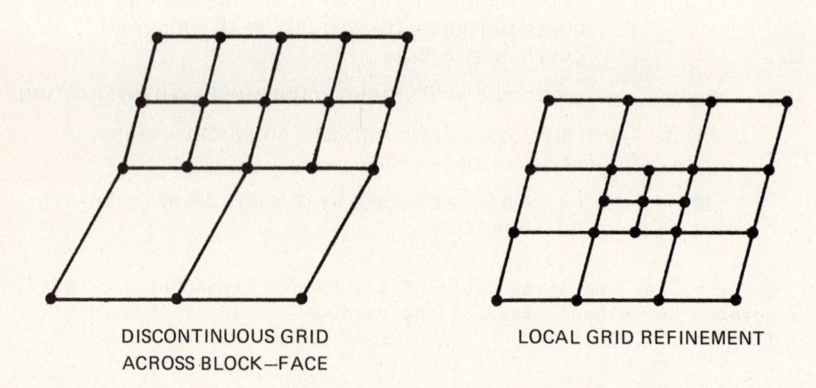

Fig. 9 Grid discontinuities (two-dimensional)

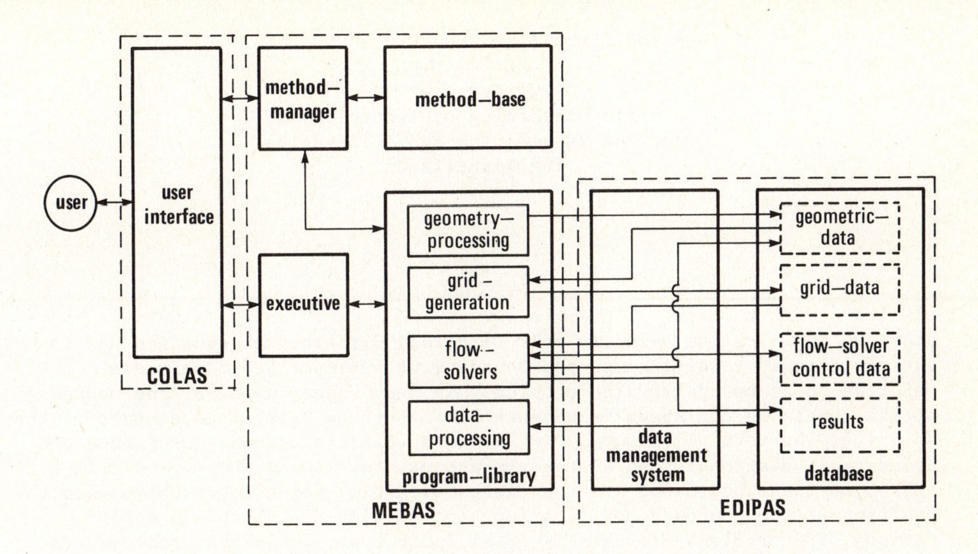

Fig. 10 General technical concept at NLR for data management, method management, and user interaction

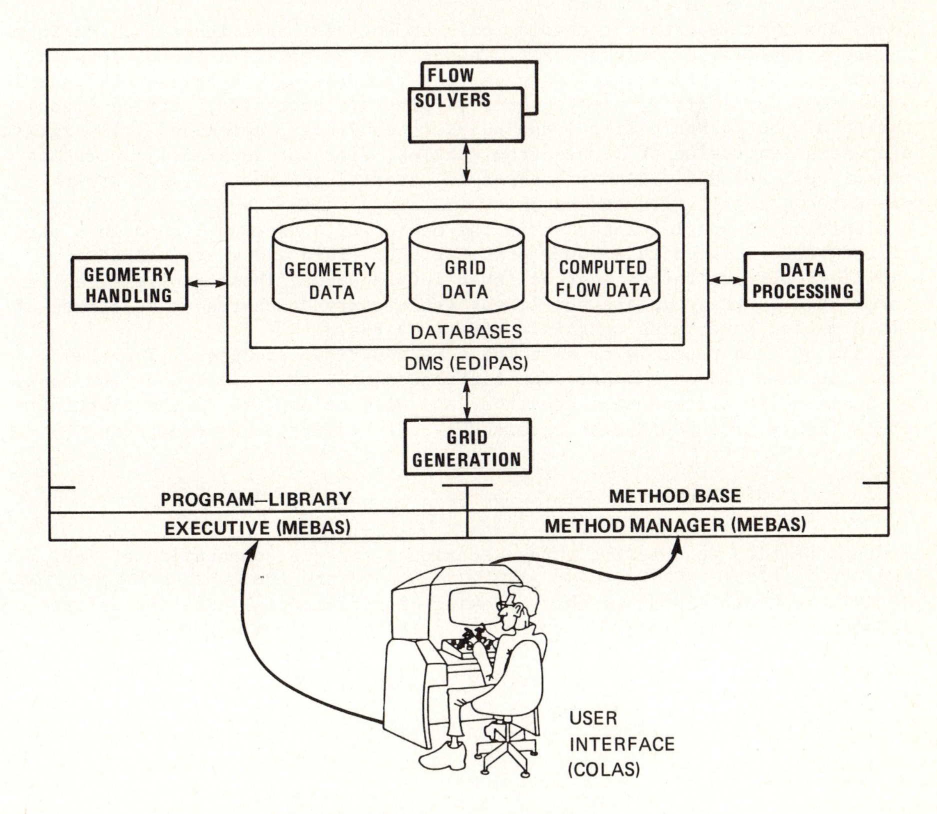

Fig. 11 Technical concept for computational aerodynamics infrastructure at NLR

ON THE COMPUTATION OF FREE BOUNDARIES

C. Cuvelier
Delft University of Technology
Dept. of Mathematics and Informatics
The Netherlands

1. INTRODUCTION

Problems in which the solution of a (partial) differential equation has to satisfy certain conditions on the boundary of a prescribed domain are referred to as boundary-value problems. In many cases, however, the boundary of the domain is not known in advance but has to be determined as part of the solution. The term *stationary free-boundary* (SFB) is commonly used when the boundary is stationary and a steady-state exists. *Moving free boundaries* (MFB), on the other hand, are associated with time-dependent problems and the position of the boundary now is a function of time and space. In applications, SFB problems are usually of elliptic type, while MFB problems are often described by parabolic equations. We refer to [1] which presents a broad and detailed account of the mathematical (both analytical and numerical) solution of FB problems.
There are many important technological and engineering-science applications in which FBs play a dominant part. This can be exemplified by FB flows in porous media associated with seepage (the dam problem), enhanced oil recovery processes, capillarity, electrochemical plating, corrosion, coating (manufacturing of photographic films) and polymer technology, metal and glass forming processes (extrusion of liquid from nozzles, fibre production), separation techniques, solidification processes in material science (crystal growth processes), lubrication and so on.
A situation of special interest is the motion of a viscous liquid in a partly filled vessel placed in a low-g gravitational field and where, moreover, capillary surface forces must be taken into account. These problems can be formulated by the complete non-linear (stationary or unsteady) equations of fluid motion on domains containing (S or M) FBs.
The aim of this paper is to describe some numerical techniques for the computation of (S and M) FBs. The techniques will be explained on the basis of a simple 1D differential equation, and will be applied to the computation of capillary free boundaries governed by the Navier-Stokes equations.

2. STATIONARY FREE BOUNDARIES

In this section we describe three techniques for the computation of SFB, viz. the *trial free-boundary method*, *Newton's method* and the *total linearization method*. As example we take the following 1D differential equation defined on a domain (interval) which is unknown a priori:

Find $u = u(x)$ and an interval $(0,\gamma)$ such that

$$\left|\begin{array}{l} -u'' = 2 \quad \text{on } (0,\gamma) \\ u(0) = 0 \\ u(\gamma) = 0 \quad u'(\gamma) = -1 \end{array}\right. \tag{2.1}$$

where ' stands for derivation with respect to x. One easily verifies that $u(x) = x(1-x)$ and $\gamma = 1$ is the solution of this problem. Notice that at the point γ two boundary conditions must be satisfied, one is necessary for the solvability of the problem if γ would be fixed, the second boundary condition is used to determine the unknown γ.

The *trial free-boundary method* is based on the following general principle. An initial estimate of the FB is assigned. The (partial) differential equation is solved within the domain fixed by the initial estimate and after disregarding one of the boundary conditions on the FB. Next, a new shape of the FB (i.e. the domain) is computed which satisfies as closely as possible the boundary condition that was relaxed. This procedure is repeated until convergence is attained.
Let us apply this procedure to problem (2.1). Let γ_0 be an initial estimate of the FB, and let us consider the following problem, now on a fixed domain

$$\left| \begin{array}{ll} -u'' = 2 & \text{on } (0,\gamma_0) \\ u(0) = 0 & \\ u(\gamma_0) = 0 \,. & \end{array} \right. \qquad (2.2)$$

The solution of this problem is given by $u = x(\gamma_0 - x)$. The second FB condition is now used to improve the estimate for γ. We define γ_1 by $u'(\gamma_1) = -1$, which gives $\gamma_1 = \frac{1}{2}(\gamma_0+1)$, or, after m+1 iterations $\gamma_{m+1} = \frac{1}{2}(\gamma_m+1)$. This is a Picard-type iteration of the form $\gamma_{m+1} = F(\gamma_m)$, with solution $\gamma = 1$ and convergence factor $F'(\gamma=1) = \frac{1}{2}$, proving that the rate of convergence of this process is linear. An other disadvantage of the trial method is that convergence is not always assured. If we had interchanged the two FB conditions, the solution of the differential equation on $(0,\gamma_0)$ would be $u = -x^2+(2\gamma_0-1)x$ and γ_1 would be given by $\gamma_1 = 2\gamma_0-1$, or generally $\gamma_{m+1} = 2\gamma_m - 1 \equiv F(\gamma_m)$, which is a divergent iteration since $F'(1) = 2 > 1$.
The trial method consists of two steps: the first step is to solve a (partial) differential equation on a fixed domain, which can be done using for instance finite difference, finite element or finite volume methods. the second step is to improve the shape of the FB.

Newton's method also involves a deforming domain but eliminates the successive iteration between the position of the FB and the field variables by introducing, in the discrete case, the position of the nodes on the FB directly as degrees of freedom. This is then coupled to a Newton method which results in the simultaneous calculation of the position of the FB and the field variables at the new nodal positions once convergence is attained.
In order to apply this method to the 1D example, let us discretize (2.1) by means of a finite difference method with an equidistant mesh h.

0 1 2 3 n n+1

0 ⟷ γ
h

$h = \gamma/n$.

This give the following system

$$
\begin{aligned}
2u_1 - u_2 &= 2h^2 \\
-u_1 + 2u_2 - u_3 &= 2h^2 \\
&\vdots \\
-u_{n+1} + 2u_n - u_{n+1} &= 2h^2 \\
u_n &= 0 \\
u_{n+1} - u_{n-1} &= -2h
\end{aligned}
$$

which is equivalent to

$$\underline{R}(\underline{u},h) = \underline{0} \tag{2.3}$$

with

$$\underline{R}(\underline{u},h) = \begin{bmatrix} 2 & -1 & & & & & -2h \\ -1 & 2 & -1 & & & & -2h \\ & -1 & 2 & -1 & & & -2h \\ & & & \ddots & & & \\ & & & -1 & 2 & -1 & -2h \\ & & & 0 & 1 & 0 & 0 \\ & & & -1 & 0 & 1 & 2 \end{bmatrix} \begin{bmatrix} u_1 \\ u_2 \\ \cdot \\ \cdot \\ \cdot \\ \cdot \\ u_{n+1} \\ h \end{bmatrix} \tag{2.3}$$

$$\underline{u} = (u_1,u_2,\ldots,u_{n+1})^T .$$

The non-linear system (2.3) is solved by Newton's method, which necessitates the computation of the Jacobian J

$$J(h) = \frac{\partial \underline{R}(\underline{u},h)}{\partial (\underline{u},h)} = \begin{bmatrix} 2 & -1 & & & & & -4h \\ -1 & 2 & -1 & & & & -4h \\ & -1 & 2 & -1 & & & -4h \\ & & & & & & \\ & & & -1 & 2 & -1 & -4h \\ & & & 0 & 1 & 0 & 0 \\ & & & -1 & 0 & 1 & 2 \end{bmatrix} .$$

Let us denote by $\{\underline{u}_0,h_0\}$ an initial approximation of the solution $\{\underline{u},h\}$, then the (m+1)th iterand satisfies

$$J(h_m) \begin{bmatrix} \underline{u}_{m+1} \\ h_{m+1} \end{bmatrix} = J(h_m) \begin{bmatrix} \underline{u}_m \\ h_m \end{bmatrix} - \underline{R}(\underline{u}_m,h_m) = - \begin{bmatrix} 2h_m^2 \\ \vdots \\ 2h_m^2 \\ 0 \end{bmatrix} . \tag{2.4}$$

When we take, for instance, n = 4, sytem (2.4) gives

$$h_{m+1} = \frac{4h_m^2}{8h_m - 1} = F(h_m)$$

which defines a Picard process with solution h = 0.25 (i.e. γ = 4*0.25=1) and convergence factor F'(0.25) = 0, so that the iteration process (2.4) converges quadratically.

In general one can say that the trial FB method is easy to program, but converges slowly. The Newton method has a quadratic rate of convergence, but suffices from the disadvantage that a complete account of the variations with respect to the FB degrees of freedom must be incorporated into the Jacobian of the system of equations. In the finite-element context, these variations, which involve integrals over a large part of the domain, have a non-local character. This means that the method does not fit into standard finite-element codes, where the coefficients in the equation of an unknown belonging to a nodal point are determined completely by contributions over the neighbouring elements.

The third method, which is termed the *total linearization method*, is much easier to implement than the Newton algorithm, while retaining its superior convergence properties. The essence of the method is that by linearizing the FB problem with respect to the FB, the influence of the unknown position of the FB can be reduced completely to the FB itself. This has great advantages for software implementation.
We write down a weak formulation of problem (2.1), where the differential equation is combined with the conditions u(0) = 0 and $u'(\gamma) = -1$.
This gives

$$\left|\begin{array}{l} \int_0^\gamma u'\phi' dx + \phi(\gamma) - 2\int_0^\gamma \phi\, dx = 0 \qquad \text{for all } \phi \text{ with } \phi(0) = 0 \\ \\ u(\gamma) = 0\,. \end{array}\right. \tag{2.5}$$

We linearize this problem with respect to γ: $\gamma = \gamma_0 + \alpha$, α small. Using the following formulas

$$\int_0^\gamma u'\phi' dx = \int_0^{\gamma_0} u'\phi' dx + \alpha u'(\gamma_0)\phi'(\gamma_0) + o(\alpha)$$

$$\phi(\gamma) = \phi(\gamma_0) + \alpha\phi'(\gamma_0) + o(\alpha)$$

$$0 = u'(\gamma) + 1 = u'(\gamma_0) + 1 + O(\alpha)$$

$$-2\int_0^\gamma \phi\, dx = -2\int_0^{\gamma_0} \phi\, dx - 2\alpha\phi(\gamma_0) + o(\alpha)$$

problem (2.5) is linearized into

$$\left|\begin{array}{l} \int_0^{\gamma_0} u'\phi' dx - 2\int_0^{\gamma_0} \phi\, dx + \phi(\gamma_0) - 2\alpha\phi(\gamma_0) = 0 \\ \\ u(\gamma_0) = \alpha\,. \end{array}\right. \tag{2.6}$$

The quantity α can be eliminated and (2.6) reduces to

$$\left| \begin{array}{l} \int_0^{\gamma_0} u'\phi' dx - 2u(\gamma_0)\phi(\gamma_0) = 2\int_0^{\gamma_0} \phi \, dx - \phi(\gamma_0) \\ \\ \text{for all } \phi \text{ with } \phi(0) = 0 \,. \end{array} \right. \tag{2.7}$$

Problem (2.7) can be considered as the weak formulation of the following boundary value problem:

$$\left| \begin{array}{ll} -u'' = 2 & \text{on } (0,\gamma_0) \\ u(0) = 0 & \\ u'(\gamma_0) - 2u(\gamma_0) = -1 \,. & \end{array} \right. \tag{2.8}$$

Let γ_0 be an approximation of the FB, then the first step in the iteration process is to solve (2.8), which gives

$$u = -x^2 + \frac{-2\gamma_0^2 + 2\gamma_0 - 1}{-2\gamma_0 + 1} x \,.$$

Next we use condition (2.6)b to find a better approximation of γ:

$$\gamma_1 = \gamma_0 + \alpha = \gamma_0 + u(\gamma_0) = \frac{\gamma_0^2}{2\gamma_0 - 1}$$

or after m+1 iterations

$$\gamma_{m+1} = \frac{\gamma_m^2}{2\gamma_m - 1} = F(\gamma_m)$$

which is a quadratic process with solution $\gamma=1$. Notice that problem (2.6) is linearized and that it can be solved by, for instance, a finite-element or a finite-difference method.
The total linerization method is closely related to an idea in [8], which consists of considering an equivalent set of boundary conditions, (depending on a parameter τ) for problem (2.1), and then to apply the trial method. The question then is, how to choose the parameter τ such that the boundary condition which is not relaxed in the trial method is stationary with respect to conormal displacements of the boundary. For problem (2.6) this is the combination

$$u'(\gamma) - 2u(\gamma) = -1$$

$$u(\gamma) = 0 \qquad \text{(relaxed condition)}.$$

An attractive method for FB computations is the method of *variational inequalities*, see [11]. This method reformulates the FB problem into a constraint optimization problem, which in some sense eliminates the unknown position of the FB. This method, however, is, as far as the author knowns, not applicable to capillary FB problems.

3. CAPILLARY FREE BOUNDARY

In this section we discuss some applications of the techniques described in the preceding section to the computation of capillary FBs in a viscous incompressible liquid with density ρ and viscosity μ. Capillarity means that we take into account the forces due to intermolecular attractions, which have a non-vanishing resultant at the FB of the liquid. The motion of the liquid (which is assumed to be stationary or static (rigid-body) in this section) is governed by the Navier-Stokes equations for the velocity $\underline{u}$ and the pressure p. On fixed walls we impose the no-slip condition and on the FB the following conditions must be satisfied

$$
\begin{aligned}
&u.\nu = 0 \\
&\sigma_\tau = 0 \\
&\sigma_\nu = \lambda\left(\frac{1}{R_1} + \frac{1}{R_2}\right) - p_a
\end{aligned}
\qquad (3.1)
$$

where σ_τ and σ_ν are the tangential and normal components of the stress tensor, where λ denotes the coefficient of surface tension, R_1 and R_2 are the principal radii of curvature, p_a is the atmospheric pressure which we put equal to zero and where ν denotes the unit normal on the FB.
When we consider the axisymmetric rigid-body rotation of a liquid partly filling a sphere (with radius R) rotating around its vertical axis with angular velocity ω, the FB is governed by the following differential equation in polar coordinates $r(\theta)$ and θ with respect to an origin 0, see Fig. 3.1.,

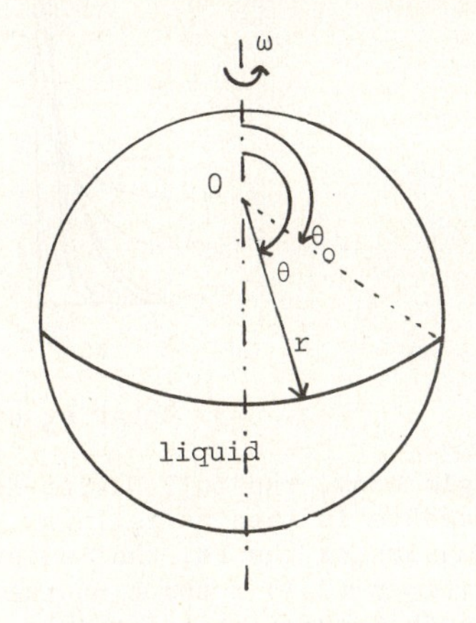

Fig. 3.1 Sphere partly filled with liquid.

$$
-\frac{2r^2+3(r')^2-rr''}{(r^2+(r')^2)^{3/2}} + \frac{r'\cot\theta}{r(r^2+(r')^2)^{\frac{1}{2}}} + \text{Bo}\; r\cos\theta - \text{We}\; r^2\sin^2\theta = c \text{ on } (\theta_0,\theta_1) \qquad (3.2)
$$

where $\text{Bo} = \rho g R^2\lambda^{-1}$ Bond number, g = acceleration of gravity, $\text{We} = \frac{1}{2}\rho\omega^2R^3\lambda^{-1}$ Weber number and where c is an unknown constant, which is related to the fact that the quantity of liquid in the vessel is given. The interval (θ_0,θ_1) is unknown (FB). Because of the axisymmetry of the problem we can take $\theta_1 \leqslant \pi$. The boundary conditions at θ_0 and θ_1 are:

if $\theta_1 < \pi$: $r(\theta_i)$ = known in terms of θ_i, $i = 1,2$

$r'(\theta_i)$ = known in terms of θ_i and the contact angle, $i=1,2$

if $\theta_1 = \pi$: $r'(\theta_1) = 0$

We have solved this problem with the trial FB method, see [4]. The differential equation (3.2) is solved on the fixed interval (θ_0,θ_1) using Newton linearization in combination with a finite-element method. At the FB points (i.e. θ_0, and θ_1 if $\theta_1 < \pi$), the Dirichlet conditions are taken into account and the Neumann condition is used to improve the position of the FB. In Fig. 3.2 we show the results of some computations with Bo=10.0 (low-gravity), filling factor 0.5, contact angle 0.05π and We = 0, 10, 20, 30, 31, 40, 50. Each solution was computed by using the solution of the previous We number as initial approximation for the trial method. Fig. 3.3 shows the results for a rotating cylinder with radius 0.5, height 2.0, filling factor 0.25, Bo=5.0, contact angle 0.05π and We = 0, 25, 40, 45, 100, 150.

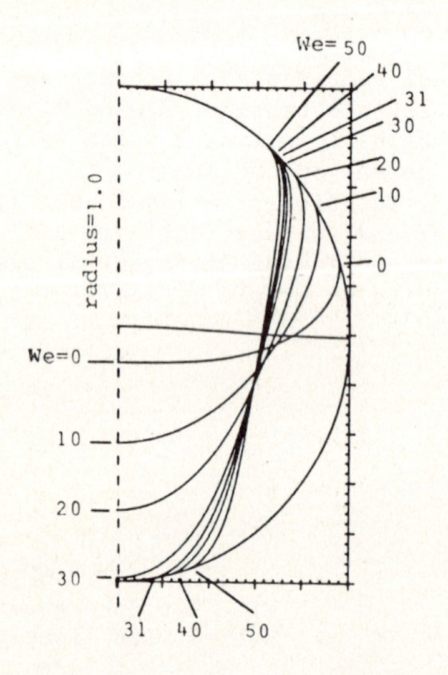

Fig. 3.2 Rotating sphere.

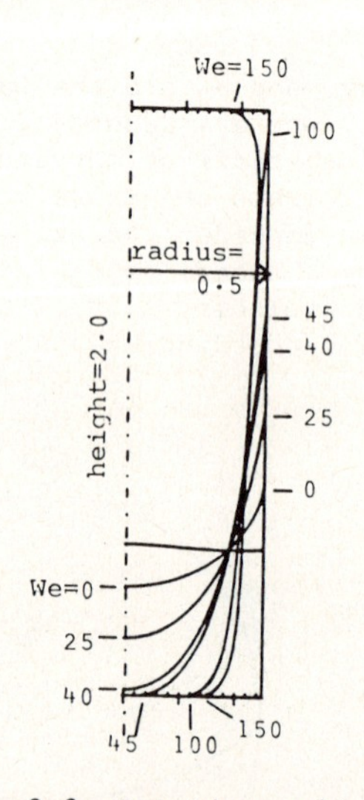

Fig. 3.3 Rotating cylinder.

When the liquid does not move like a rigid body, the full Navier-Stokes equations must be solved. The trial procedure is then as follows. Within the domain determined by an initial approximation of the FB, the Navier-Stokes equations are solved with boundary conditions (3.1) a and b on the FB. Once a solution $\{\underline{u},p\}$ is obtained, we calculate the normal stress $\sigma_\nu \equiv \sigma_\nu(\underline{u},p)$ and we solve the differential equation for the normal stress (3.1)c to obtain a better approximation of the FB.
The Navier-Stokes equations can be solved by triangular or quadrilateral finite elements in combination with a penalty method for the incompressibility constraints. For the velocity we have chosen P_2^+ triangular or Q_2 quadrilateral elements together with an element-wise linear non-continuous approximation for the pressure. The velocities in the centroid of the elements have been eliminated with the aid of the discrete continuity equations. The differential equation for the FB is solved numerically using quadratic 1D elements. For an introduction to the finite-element analysis of the Navier-Stokes equations, we refer to [6] in which also further references

are given concerning the computation of FBs. For the 2D situation of a liquid in an open rectangular vessel, as depicted in Fig. 3.4.,

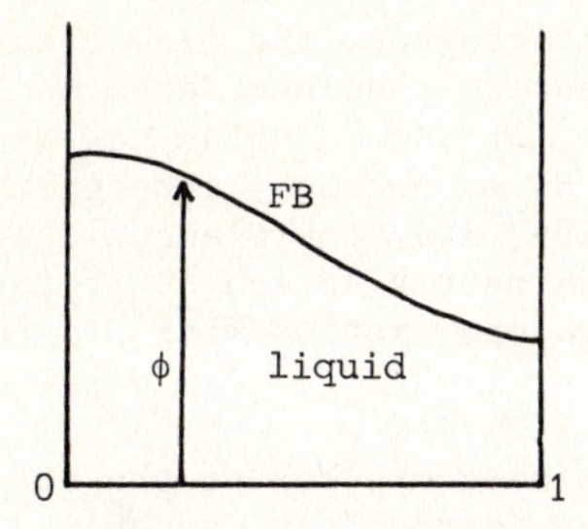

Fig. 3.4 Liquid in 2D rectangular vessel.

the normal stress differential equation reduces to

$$\left|\begin{array}{ll} \lambda\left(\dfrac{\phi'}{\{1+|\phi'|^2\}^{\frac{1}{2}}}\right)' = -p + 2\mu\,\dfrac{\partial \underline{u}.\nu}{\partial \nu} & \text{on } (0,1) \\ \int_0^1 \phi\, dx = \text{volume of liquid} & \\ \phi'(0),\ \phi'(1) \text{ given} \quad (\text{contact angles}) & \end{array}\right. \qquad (3.3)$$

where ϕ represents the shape of the FB. Notice that, since the Navier-Stokes equations are solved with boundary conditions $\sigma_\tau=0$ and $\underline{u}.\nu=0$ on the FB, the pressure is determined up to an additive constant. The (non-linear) Navier-Stokes equations and the (non-linear) FB equation (3.3) can be solved most efficiently by performing a Newton linearization. Nevertheless, the FB algorithm itself, being of trial-method type, has a linear rate of convergence. We refer to [2], [6].

Both the trial method as well as the Newton method have been applied to the computation of a 2D FB arising from the steady thermo-capillary flow of a viscous incompressible fluid, see [5], [6]. The problem is considered in the context of open boat crystal growth technique and can be modelled by the Navier-Stokes equations coupled with the heat equation in a rectangular container. There exist two types of couplings between the Navier-Stokes equations and the heat equation. First, the Boussinesq approximation is used to handle the temperature dependence of the density. On the other hand, the coefficient of surface tension λ is considered as depending linearly on the temperature. This gives rise to a FB condition where the tangential stress is inversely proportional to the gradient of the temperature on the FB. As we mentioned before, the trial method is easy to program, but has linear convergence properties, while the Newton method converges quadratically. However, the programming is difficult, because of the computation of the Jacobian. The set of equations can be written in terms of the residuals $\underline{R}(\underline{u},\underline{p},\underline{\phi})$. In the Jacobian, the components

$$\frac{\partial \underline{R}}{\partial \underline{u}} \quad \text{and} \quad \frac{\partial \underline{R}}{\partial \underline{p}}$$

arise in any fixed boundary flow problem and there is no difficulty in calculating them. The difficulty in FB problems is the accurate calculation of the components of

$$\frac{\partial \underline{R}}{\partial \underline{\phi}}$$

i.e. sensitivities of all the residuals throughout the flow domain to the shape of the FB. The residuals are expressed by volume, area and line integrals. The evaluation of derivatives for the Jacobian matrix is very tough, because the residuals all depend on $\underline{\phi}$, not only through the integrands but also through the limits of integration, i.e. the element shapes. The costs of the construction of the Jacobian matrix at ech FB iteration can be reduced by using, for instance, a quasi-Newton method with Broyden update. See for instance [6].

The total linearization method (TLM) has been applied to the die-swell problem, which is an important phenomenon in rheology, see [9], [10]. The die-swell problem is an extrusion problem of a viscous incompressible jet from a die into an inviscid fluid.
Because FB problems are non-linear, the equations must be linearized with respect to the position of the FB, velocity and pressure. The basic idea of the TLM is to linearize the continuous form of the weak formulation. This is in contrast with Newton's method which performs a linearization of the discrete form of the FB problem. The second essential point is to take account of the boundary conditions at the FB in order to simplify the linearization of the weak formulation. In this way the influence of the free boundary can be reduced to boundary integrals only, which simplifies the computation of the Jacobian. In the finite element context, the variations of the residuals with respect to the position parameters of the FB now have a local character, in the sense that the coefficients in the equation for an unknown belonging to a nodal point are determined by the contributions of neighbouring elements. Although it is not yet clear whether or not the TLM has a quadratic rate of convergence, it is superlinear. Its major advantage over the Newton method is the restriction of the geometrical unknowns to the FB, which reduces the profile of the stiffness matrix. A second advantage of the TLM lies in the ease of implementation on the computer.

4. MOVING FREE BOUNDARIES

In this section we consider a typical moving boundary (MB) problem of Stefan type and we mention some techniques, like front-tracking, fixed domain techniques, which can be used to solve this type of problems. Fundamental in such problems is the presence of an interface which moves with time and has to be determined in the course of the computation. Consider, for instance, a sheet of ice [0,1] initially at melting temperature equal to zero. The temperature at the surface is raised to a temperature T=1 at time t=0. The interface on which melting occurs moves into the ice and separates a region of water with temperature T>0 and a region of ice with T=0. So, the problem is to find a temperature distribution T(x,t) such that the heat conduction equation is satisfied in the water region $(0,\gamma(t))$ with heat conduction coefficient κ:

$$T_t = \kappa T_{xx} \qquad x \in (0,\gamma(t)) \; , \qquad t > 0 \tag{4.1}$$

subject to the following boundary condition at x=0:

$$T(0,t) = 1 \; . \tag{4.2}$$

The initial conditions are given by

$$T(x,0) = 0 \;, \qquad x \in [0,1] \;. \tag{4.3}$$

At the interface $x=\gamma(t)$, two conditions are needed. One to provide a second boundary condition for the heat equation and a second one to determine the position of the interface, we impose

$$\begin{aligned} &T(x=\gamma(t),t) = 0 && \text{continuity of temperature} \\ &-\kappa\, T_x = \ell\, \gamma_t && \text{heat-flux balance} \end{aligned} \tag{4.4}$$

where the second condition expresses that the heat flow from water to ice is transformed into latent heat (ℓ = latent heat coefficient).
The techniques used for the numerical solution of MB problems either belong to the class of front-tracking methods, where the position of the MB is predicted along with the solution of the differential equation, or they belong to the class of enthalpy and variational-inequalities methods, where a reformulation of the problem is solved in a fixed domain and the position of the MB is determined a posteriori.

Front-tracking method compute at each time step the position of the MB. When the solution is computed at points on a fixed grid in space and time, the boundary will in general be between grid points at any time. Therefore, special formulas are needed to handle terms like u_x and γ_t, as well as the (partial) differential equation itself, in the neighbourhood of the MB. For example, an explicit scheme for the new position of the MB at time $t=n\Delta t$ is

$$\ell\{\gamma(t_{n+1}) - \gamma(t_n)\} = -\kappa\, \delta_x T^n$$

where $\delta_x T$ denotes the finite-difference or finite-element approximation for T_x at $x=\gamma(t_n)$. If an implicit approximation such as

$$\ell\{\gamma(t_{n+1}) - \gamma(t_n)\} = -\frac{\kappa}{2}\{\delta_x T^{n+1} + \delta_x T^n\}$$

was used, an iterative scheme must be employed since both T^{n+1} and $\gamma(t_{n+1})$ are unknown. This makes the implicit method very expansive, even if just two iterations (predictor-corrector) are used. Often it is better to approximate $\delta_x T^n$ accurately, by choosing a refined grid in the neighbourhood of the MB.
An alternative way to apply a front-tracking method is to define a moving grid in such a way that the MB is always on a grid line. The main problem still remains the computation of derivatives near the MB in order to satisfy the Stefan conditions.

A second approach to MBs is formed by the *fixed-domain methods*, like the *enthalpy method* and the formulation as a *variational inequality*. In this approach, the problem is reformulated on a fixed domain and the Stefan condition is implicitly satisfied. The position of the moving boundary can be calculated a posteriori. Let us apply the enthalpy method to problem (4.1),...(4.4). The procedure is to introduce the enthalpy $H(T)$, defined by

$$H(T) = \begin{cases} 0 & \text{if } T < 0 \qquad (0 = \text{melting temperature}) \\ T+\ell & \text{if } T > 0 \end{cases}$$

$$0 \leq H(T) \leq \ell \quad \text{if} \quad T = 0 \,.$$

Equation (4.1) is now replaced by

$$H_t = \kappa T_{xx} \qquad \text{on } (0,1) \,!$$

where T satisfies the original boundary conditions at x = 0 and x = 1, with initial conditions at t = 0. This differential equation is not defined in the classical sense, since H is discontinuous. But it is meaningful in a weak sense and it can be proved that the new formulation is equivalent to the old one. The position of the moving boundary is recovered afterwards by

$$\gamma(t) = \{x \in [0,1] \mid 0 < H(T(x,t)) < \ell\} .$$

Application of the enthalpy method to the moving interface of two inmiscible viscous fluids can be found in [7]

The *method of variational inequalities* reformulates the problem in terms of the freezing index w:

$$w(x,t) = \int_0^t T(x,\tau)\, d\tau .$$

It can be shown that w satisfies the following variational inequality

$$\left| \begin{array}{ll} w_t - \kappa w_{xx} \geq -\ell \;, \quad w \geq 0 & \\ & \text{on } (0,1) \\ (w_t - \kappa w_{xx}) w = 0 & \\ w(0,t) = t \;, \quad w(1,t) = 0 & \\ w(x,0) = 0 \,. & \end{array} \right.$$

Discrete (finite-difference or finite-element) approximations of this problem are equivalent to quadratic programming problems of the type

$$\text{minimize} \quad \tfrac{1}{2}\underline{w}^T A \,\underline{w} + \underline{b}^T \underline{w} \qquad \text{such that} \quad \underline{w} \geq 0$$

which can be solved using, for instance, projected over-relaxation or conjugate-gradients methods.
In [12] an application can be found of the moving-grid method and the variational-inequality method to a chemical etching problem. Many extensions and generalizations of the problems and methods described above exist, we refer to [1], [11].

A different approach can be studied when the MB can be considered as a *small perturbation* of a static (or stationary) situation. In this case the MB is linearized with respect to the static FB and the differential equation can be solved on a fixed domain. A first order approximation of the deviation of the MB from the static FB can be computed afterwards using one of the MB conditions.
In [3], the small perturbation approach has been applied to a 2D moving capillary FB problem in a rectangular vessel, see Fig. 3.4. The three MB conditions

$$\left| \begin{array}{l} \phi_t = \{1 + |\phi'|^2\}^{\frac{1}{2}} \underline{u}\cdot\nu \\ \\ \sigma_\tau = 0 \qquad \sigma_\nu = \lambda \left(\dfrac{\phi'}{\{1+|\phi'|^2\}^{\frac{1}{2}}} \right)' \end{array} \right. \qquad (4.5)$$

reduce, after linearization with respect to the static FB ϕ_0, to

$$\Lambda_t = \underset{\sim}{u}.\nu \qquad (4.6)$$

$$\sigma_\tau = 0 \qquad \sigma_\nu = \frac{\partial^2\Lambda}{\partial s^2} + \Lambda \{(\frac{\phi'}{\{1+|\phi'|^2\}^{\frac{1}{2}}})'\}^2$$

where Λ denotes the first order deviation of ϕ from ϕ_0 along the normal on ϕ_0 and s denotes the curve-length on the FB ϕ_0.
The Navier-Stokes equations can now be solved on a fixed domain given by the liquid in its static situation, using the boundary conditions (4.6)b en c. The first order deviation of the MB from the static FB can be computed from (4.6)a afterwards.

REMARKS

Subject of our current research is the study of fluid-structure interactions as well as the application of numerical methods and the construction of a computer code for this type of problems. Our aim is to carry out frequency analysis of fluid-structure systems, where we have in mind applications in the field of aerospace sciences, such as stabilisation and control of satellite motions.

REFERENCES

[1] J. Crank: Free and moving boundary problems. Oxford Science Publications, 1984.
[2] C. Cuvelier: Report 82-09, Delft University of Technology 1982.
[3] C. Cuvelier: Comp. Meth. Appl. Mech. Engng, 48, 1985, p.45-80.
[4] C. Cuvelier: to appear.
[5] C. Cuvelier, J.M. Driessen: J. Fluid Mech. 169, 1986, p 1-26.
[6] C. Cuvelier, A. Segal, A.A. van Steenhoven: Finite-element analysis and Navier-Stokes equations, Reidel Publ. Comp., 1986.
[7] A. Dervieux: Rapports de Recherche INRIA, 67, 68 (1981).
[8] P.R. Garabedian: Bull. Amer. Math. Soc. 62, 1956, p. 219-235.
[9] N. Kruyt: Master's thesis, Delft University of Technology, 1985.
[10] N. Kruyt, C. Cuvelier, A. Segal, J. v.d. Zanden: Submitted to Int. J. Num. Meth. Fluids.
[11] R.Glowinski, J.L. Lions, R. Trémolières: Numerical analysis of variational inequalities. North-Holland, 1981.
[12] C. Vuik C. Cuvelier: J. Comp. Physics 59, 1985, p. 247-263.

THERMAL HYDRAULIC MODELLING STUDIES ON HEAT EXCHANGING COMPONENTS

D. van Essen; G. Küpers; H. Mes.
B.V. Neratoom, P.O.Box 41155, 1009 ED Amsterdam,
The Netherlands

ABSTRACT

To perform 2-D and 3-D thermal hydraulic calculations on heat exchanging components for the liquid metal fast breeder reactor SNR-300 in Kalkar several computational models have been developed using the PHOENICS code. Calculations have been performed to predict thermal hydraulic behaviour under steady state as well as under transient conditions. In this paper the computational and validation results will be discussed for an 85 MWth intermediate heat exchanger and a 17 MWth steam generator operating under once through conditions. Both these components, prototypes for the SNR-300, were full scale tested, in the Sodium Component Test Facility (SCTF) in Hengelo, The Netherlands. It became apparent that the models developed are able to predict the most relevant characteristics of the thermal hydraulic behaviour of both the components. In the validation procedure it appeared that an excellent agreement between the measurements and the calculation results could be established.

INTRODUCTION

This paper deals with the multi-dimensional thermal hydraulic computational modelling of the intermediate heat exchanger (sodium-sodium) and the steam generator (sodium-water/steam) of the SNR-300 in Kalkar. The SNR-300 is a liquid metal fast breeder reactor built in common effort by Belgium, Germany and The Netherlands. Erection has been completed and the first criticality is foreseen in 1987.
In a nuclear power station like the SNR-300 the large heat exchanging components play an important role in reliability as well as in safety. During the engineering phase it is, therefore, necessary to know, as realistically as possible, the behaviour of these components under normal plant and safety related conditions. This was achieved by performing extensive full-scale tests of the components in the Sodium Component Test Facility in Hengelo, The Netherlands.
In parallel to the experiments, computational engineering methods were developed and validated against the experiments. Initially only lumped parameters and 1-D models were set up to predict the general thermal hydraulic behaviour of the components. Afterwards multi-dimensional models were developed to perform detailed analysis.
In this paper much attention will be paid to describe the real components and the problems to model these components for 2-D and 3-D studies. This is done to highlight the place and value of thermal hydraulic modelling in the whole process of design, detailed engineering and testing of components for nuclear power plants.

Detailed numerical modelling started with the intermediate heat exchanger. Although the prototype intermediate heat exchanger was well instrumented, it appeared that detailed 2-D and 3-D thermal hydraulic calculation results were needed for the following reasons:

- To assure the structural integrity of the component by the structural mechanical analysis, very detailed thermal hydraulic boundary conditions were needed, which could not be derived from the experimental results.
- In the safety analysis thermal hydraulic boundary conditions were needed which could not be derived from the experiments without influencing the structural integrity of the component and/or test facility.
- During the experiments some thermal hydraulic features were measured which could only be explained in a qualitative way; these features were the most important boundary conditions for the model and will be discussed further on in this paper.

After sucessfully modelling the intermediate heat exchanger with the PHOENICS code, it was decided to extend the 2-D and 3-D modelling to steam generators. In the validation procedure use has been made of the experiments carried out in the well instrumented prototype superheater, which was operated under once through conditions. The important boundary conditions derived from the experiments for a steam generator model will be discussed later.

THE 50 MWth SODIUM COMPONENT TEST FACILITY

The purpose of the 50 MWth facility was the testing of components for sodium systems under conditions representative for liquid metal fast breeder reactors, like the SNR-300 (ref./1/). A basic diagram of the test facility is given in figure 1, from which it can be seen that the circuit consisted of a sodium loop and a water/steam loop. The sodium is heated in the gas furnace (consisting of three natural-gas-fired heaters with a total capacity of 58 MWth) and pumped through the piping system by means of a sodium pump with a capacity up to 450 kg/s. In the case of testing the intermediate heat exchanger, the sodium flows from the furnace to the primary side of the component, where heat is transferred to the secondary side of the component. When the sodium has left the primary side of the heat exchanger it is forced to flow to the secondary side of the heat exchanger, where the sodium is heated up. In this recuperative arrangement it was possible to test the intermediate heat exchanger on a full scale with a capacity up to 85 MWth.
By means of several by-passes with control valves, an air-cooler and a steam generator the required flows and temperatures could be achieved.
In the case of testing the steam generator, the intermediate heat exchanger is by-passed. The sodium then flows directly to the steam generator where the heat is transferred to the water/steam side. In the water/steam loop the feedwater pump

(with a maximum capacity of 36 kg/s at 26,5 MPa) forces the water through the preheater and the steam generator, where it is transformed into steam. By means of a system of pressure-reducing valves, an air-cooled condenser and a de-aerator, the steam is transformed into water at rated pressure, temperature and quality.

DESCRIPTION OF THE COMPONENTS TESTED

In the intermediate heat exchanger heat from the primary sodium circuit is transferred to the secondary sodium circuit. In the steam generator the heat from the secondary sodium circuit is transferred to the water/steam system where steam of 500°C and 18 MPa is generated.
The two components have in common that the bundle consists of long slender straight tubes connected to the tube plates. As a consequence of the choice of the straight tube design the main problems dealt with, and thus to be solved, are caused by the non-uniform temperature distributions in the different parts of the components, leading to different thermal expansions of component parts, possibly initiating buckling of tubes.
The different thermal expansions within the bundle area are the results of two factors:

- different geometry of the shell side flow channels in the inner and outer regions of the bundle.
- different flow paths of the shell-side sodium in the bundle (the sodium in the outer region has the shortest flow path; the sodium in the inner region the longest with additional heat transfer through cross flow).

The intermediate heat exchanger (see figure 2)

The primary sodium enters the intermediate heat exchanger via the primary inlet nozzle and flows into the inlet annulus, where it is distributed along the circumference. In the annulus, formed between the inner and outer shells, the sodium passes three concentric, uniformly perforated, annular rings and flows into the shell-side area of the bundle via the uniformly perforated upper part of the inner shell.
Within the bundle, the primary sodium flows vertically downwards along the tubes exchanging heat to the secondary sodium flowing upwards inside the tubes. The tube support grids are designed in such a way that the primary flow is essentially parallel to the tube orientation. The primary sodium leaves the bundle through the uniformly perforated openings at the bottom section of the inner shell. It then flows upwards into the outlet annulus and exits via the outlet nozzle. The inlet and outlet annuli are separated from each other at approximately the axial midpoint of the bundle.
The tubes are arranged in a triangular pitch, supported by grids and at the upper and lower ends connected to tube-plates. These plates are mounted on a central tube and at the shell-side protected from primary sodium thermal shocks by thermal

shields. Much care has been taken to minimize the by-pass flows along the outer and inner peripheries of the tube bundle. The boundary at the outer edge consitutes approximately a dodecagon. Here the primary sodium is guided by a flow shroud, consisting of thin plates, which are flat at those sides where the tubes are in line and which are corrugated at those sides where the tubes are staggered. The gap between the tubes of the inner row and the central tube is provided with flow resistance plates which are fixed to the central tube.
The concept requires that the bundle should be removable for repair, if needed. To that end, the bundle can move freely through the inner shell. A labyrinth ring has been placed at the periphery of each support grid in order to prevent undesirable by-pass flows in the space between flow shroud and inner shell. These rings have only little clearance in the inner shell, and as a result the by-pass flow is kept within acceptable limits.
The secondary sodium enters the component at the top and flows downwards through a central downcomer into a semi-elliptical tube side inlet plenum. A gas filled gap between the downcomer and the central tube prevents thermal interaction between the secondary sodium in the downcomer and primary sodium in the bundle. The secondary sodium enters the bundle via an orifice plate, that is designed to provide 43% more flow to the 120 outermost tubes of the bundle than to the rest of the tubes of the bundle (see refs. /2/, /3/).
The secondary sodium is collected in an annular header and leaves the component via the secondary outlet nozzle.
The main dimensions are listed in table 1.

<u>The steam generator</u> (see figure 3)

The sodium enters the component via the sodium inlet nozzle and flows into the inlet plenum, an annulus between the flow shroud and outer shell. In the inlet plenum the sodium is distributed along the circumference and flows upwards. The top part of the flow shroud is uniformly perforated, which allows the sodium to enter the shell side area of the bundle. Within the bundle the sodium flows vertically downwards transferring heat to the water/steam which is flowing upwards inside the tubes.
The sodium leaves the shell side area of the bundle via a uniformly perforated part of the flow shroud at the bottom and flows into the outlet annulus. In the outlet annulus the sodium flows upwards and leaves the component via the sodium outlet nozzle.
At the top and the bottom the tubes are connected into tube-plates, which are, at the shell side, protected against thermal shocks by thermal shields. The tubes are arranged in a triangular pitch, supported by grids and surrounded by a circular flow shroud. Thermal expansion between the outer shell and the tubes of the bundle is compensated by a bellows construction in the outer shell.
The water enters the component via the feedwater inlet nozzle at the bottom part of the component and flows into the feed-

water inlet header. In this header the water is distributed over the tubes. In the tubes the water flows upwards and is heated up by the sodium flowing around the tubes. Depending on the process conditions the water is transformed into wet or superheated steam. At the top of the component the steam out of the tubes is collected in a steam header and leaves the component via the steam outlet nozzle. The main dimensions of the component are listed in table 2.

INSTRUMENTATION

Both the components were well instrumented in order to verify the proper functioning of the components and to provide details of the thermal hydraulic behaviour.
Besides measuring the overall in and outlet conditions, thermocouples were placed at the shell side in order to determine axial, radial and circumferential temperature profiles at several levels and radial positions. At the tube side, radial and circumferential outlet profiles could be determined. Further details of the instrumentation can be found in refs. /1/, /2/, /4/, /5/.

EXPERIMENTS PERFORMED

Besides unsteady state experiments, like scrams, load changes, start-up and shut-down procedures, steady state experiments were performed under the following process conditions:

For the intermediate heat exchanger:

Primary sodium inlet temperature	: 400-550°C
Difference between primary and secondary sodium inlet temparature	: 0-215 K
Primary mass flow rate	: 10-400 kg/s
Ratio between primary and secondary massflow	: 0.7-1.4

For the steam generator:

Sodium inlet temperature	: 350-500°C
Sodium mass flow rate	: 60-220 kg/s
Water inlet temperature	: 220-300°C
Water mass flow rate	: 3-16 kg/s
Steam outlet temperature	: 280-450°C
Steam outlet pressure	: 6-18 MPa

EXPERIMENTAL RESULTS

To discuss all the experiments performed and the measurements obtained is beyond the scope of this paper. Only a few phenomena derived from the experiments will be discussed, leading to the important boundary conditions for both the computational models.

For the intermediate heat exchanger

Heat transfer in the component is governed by the ratio of the primary to the secondary mass flow and by the absolute primary mass flow rate. From the experiments it appeared that the radial temperature difference at the tube outlet side was strongly dependant on both these parameters (see figure 4). From this figure it can be seen that:

- the temperature difference increases when the mass flow ratio decreases,
- the temperature difference increases when the primary mass flow rate decreases,
- below a certain value of the primary mass flow rate the temperature difference decreases by decreasing the primary mass flow rate.

Calculation results with 1-D or 2-D computer codes in which free convection was neglected showed the increase of the temperature difference with decreasing mass flow rate and/or mass flow ratio but did not show the decrease at very low mass flow rates.
Figure 5 gives the measured temperatures in the case of an experiment whereby the mass flow ratio, as well as the primary mass flow rate were very low.
From this figure it can be seen that:

- heat is transferred in the upper part of the bundle only,
- the radial temperature difference at the top of the bundle is rather small,
- the temperature difference between the sodium inlet temperature near the inlet nozzle and the sodium entering the bundle is rather large,
- there is a large temperature difference between the temperatures in the inlet plenum and the by-pass area between the flow shroud and inner shell.

Now it can be assumed that the difference between the temperatures near A and B is very large, so heat will be transferred from the inlet plenum to the by-pass area. This means that the temperature near C is lower than the temperature near A and so the heat transferred near C and D will be less. Consequently, it can be assumed that the temperatures near D will be lower than near B. In this case free convection must be taken into account and in particular situations it could occur that sodium is flowing upwards from B to D, so cold sodium from the bottom part of the component will flow upwards and will mix with the hot sodium entering the bundle region.
This was leading to a very important boundary condition for the computational model to be developed e.g. besides forced convection phenomena it has to deal with free convection phenomena; i.c. a coupled solution of the momentum and energy equations. In figure 6 the calculation results show that free convection swirls are occurring in the area between flow shroud and inner shell.

More validation results for this case and other cases can be found in reference /6/.

For the steam generator.

In steam generators one has to deal with completely different phenomena than in intermediate heat exchangers. Instead of only one heat transfer phenomenon on the tube side, one has to deal with a number of heat transfer phenomena due to a number of flow regimes (preheating, boiling, superheating). From figure 7, indicating the steam outlet temperature in radial direction, it can be seen that much higher temperatures were measured at the outside area of the bundle than in the middle and centre of the bundle. The large radial temperature differences are caused by the geometrical configuration of the outer bundle area. As mentioned before, the tubes in the bundle are arranged in a triangular pitch and surrounded by a circular flow shroud, leading to large sodium by-pass flows in the outer bundle region.
Comparison of the conditions in the outside area of the bundle with those in the other parts of the bundle leads to the following:

- in the outside area more sodium is available per tube,
- in the outside area more heat will be transferred to the water/steam side of the tube,
- in the outside area more water will be transformed into steam, leading to longer steam columns in the tubes,
- the pressure drop over the tubes in the outside area tends to be higher, because the contribution to the total pressure drop due to friction is larger than that due to gravity (especially in the case of high water/steam velocities as a result of large feedwater flow rates).

However, for the component discussed in this paper, it has to be taken into account that the tubes are connected to common water and steam headers. This means that the pressure drop over each tube has to be identical, requiring a non-uniform feedwater flow distribution in the tubes.
This result implies the most important boundary condition for modelling a steam generator, e.g. the feedwater flow distribution through the tubes has to be calculated by an iterative procedure over the total pressure drop.
The calculated feedwater flow distribution is given in figure 8, while for the validation procedure the accompanying steam outlet temperatures are given in figure 7.

THE MATHEMATICAL MODEL

The 2-D and 3-D finite difference representations of the components described before are based on the general purpose PHOENICS program (ref. /7/). This program is composed of three main elements: Earth, Satellite and Ground.

The main element Earth contains the algorithms for the formulation and solution of the discretised forms of the differential equations governing conservation of the dependant variables. The elements Satellite and Ground contain the problem specifications such as geometry, thermo-hydraulic resistance formulae, fluid properties and boundary conditions for the components to be calculated.
The model uses the well-tried concepts of volume porosity, surface permeability and distributed resistance to represent the influence of flow blockages and resistance to fluid motion with the finite subdomains into which the flow geometry is subdivided. These, together with details of the differencing scheme and solution method embodied in the PHOENICS code, have been described elsewhere (ref. /8/).

THE COMPONENT BOUNDED MODELS

With the experience gained while setting up and performing calculations with the model for the intermediate heat exchanger, the steam generator model has been set up keeping in mind special features such as: operational flexibility and applicability of the model for problems of a more general nature. This resulted in four important features in the steam generator model.

a. Generalisation of geometrical input data.

The set-up of the geometry input specification for the steam generator model has been done in such a way that it facilitates incorporation of additional internal features and automates the finite domain grid generation process.
In the intermediate heat exchanger model the user has to prescribe the actual nodal locations, while in the steam generator model he only has to put in the spatial locations of the principal internal features together with the desired number of mesh points between adjacent features. The input sequences in the latter model then automate the generation of the appropriate grid mesh from this specification. This enables the user to use the model as a design aid and to perform a sensitivity analysis on the grid-mesh in a very simple way. Furthermore, the program performs a number of consistency checks of the input, to assist the user in simulating a particular geometry.

b. Modular implementation of physical model inputs.

Where needed, physical model inputs were placed in the model for the intermediate heat exchanger.
In order to make the model as general as possible, the physical model inputs to the steam generator program have been set up in modular form to simplify the substitution of alternative correlation sets when performing parameter studies.

The physical model input modules include sequences for specifications of shell-fluid and tube-fluid property data, turbulence models, correlations for heat transfer between tubes and shell- and tube-fluids and correlations representing the friction losses encountered by the fluids passing over and within the tubes of the bundle.

c. Features of the tube-fluid treatment.

In an intermediate heat exchanger one does not have to deal with boiling phenomena and large changes in the heat flux because of the sodium being on the shell side as well as on the tube side.
As a result the grid mesh can be equal for both sides.
In a steam generator, on the other hand, different liquids are present at the shell side and the tube side and phase changes occur within the tubes together with large changes in the heat flux near the DNB zone. During transients this DNB zone moves along the tube wall.
In the steam generator model the computation of the water side thermo-hydraulic effects is characterised by two important features:

- The use of separate mesh distributions for shell side and tube side analysis, with appropriate boundary value matching. It is now possible to use a finer mesh on the tube side than on the shell side, particularly in vertical direction. Because of this it is possible to predict more accurately the spatial variations in the location of the DNB zone and to track shifts in the DNB zone during transient simulations.
- A built-in iterative scheme affording stabilised recomputations of spatial variations in tube-fluid flow rate. This allows reconciliation of the imposed uniform pressure drop condition over all the tubes with the three-dimensional heat flux calculation from the shell fluid analysis. This updating of the tube-fluid flow rate distribution takes place in a numerically stable manner as the overall calculation of shell and tube side thermo-hydraulic phenomena proceeds. This does not appear to be a feature of many other published models which are constrained to operate with uniform or prescribed tube-side distributions.

d. Flexibility of boundary condition specification.

In all circumstances the intermediate heat exchanger is operating under process driven boundary conditions and, normally, the steam generator is operating under turbine driven boundary conditions.
In some cases the steam generator is operating under process driven boundary conditions. This means that this dual capability had to be built in into the steam generator model. For the validation procedure the process driven possibility is used because of the shorter computation time.

VALIDATION RESULTS

A complete validation procedure will feature comparison of 2-D and 3-D calculation results against measurements for steady state full load and low load conditions as well as for unsteady state conditions. Furthermore, a sensitivity analysis for several parameters, such as turbulence models, set of heat transfer correlations, etc. will be performed.
The validation results for the intermediate heat exchanger are given in ref. /6/, while an important calculated feature, the influence of free convection, is given in figure 6.
The validation procedure of the steam generator has not yet been finished. The first calculation results for full load conditions are given in the figures 7 and 8.

CONCLUSIONS

From the results presented it appeared that with the models developed for both the intermediate heat exchanger and the steam generator it is possible to describe the thermal hydraulic behaviour of both the components very accurately. It is now possible to perform parameter and design studies for intermediate heat exchangers as well as for steam generators in a very easy way.

AKNOWLEDGEMENTS

The studies in this paper were sponsored by the Dutch Ministry of Economic Affairs, as a part of the Sodium Technology Development Program. The set-up of both the models has been done in close cooperation with Cham Ltd., the originators of the PHOENICS code.

REFERENCES

/1/ Ludwig, P.W.P.H., Hus, B.W.: "Some results of the 50 MW straight tube steam generator tests in the TNO 50 MW SCTF at Hengelo", Summary report IAEA Study Group Meeting on Steam Generators for LFMBR's, Bensberg, October 14-17, (1974).

/2/ Leeuwen, N.J.M. van, Westerweele, W.J.:"Expieriences during testing of a prototype intermediate heat exchanger for SNR-300". Proceedings of Second International Conference on Liquid Metal Tenology in Energy Production, Richland, Washington, April (1980).

/3/ Hirs, G.G., Horst, J.F.M. ter: "Solving thermal hydraulic problems in heat exchangers by flow redistribution" ASME-WAM (1982), 82-WA/NE-2.

/4/ Brasz, J, Essen, D. van: "Experimental determination of density-wave oscillations in full-scale sodium-heated steam generators". Proceedings of the Second International Conference on Boiler Dynamics and Control in Nuclear Power Stations, BNS, Bournemouth, October 23-25, (1979).

/5/ Essen, D. van: "Experimental results of the 85 MWth SNR-300 intermediate heat exchanger". ASME publucation, HTD-Vol. 51, pp. 1-7, ASME-WAM, (1985).

/6/ Kirkcaldy, D., Phelps, P.J., Essen, D. van: "Phoenics code thermal hydraulic analysis of the SNR-300 IHX". ASME Publication, HTD-Vol. 51, pp. 9-16, ASME-WAM, (1985).

/7/ Spalding, D.B.: "A general purpose program for multidimensional one- and two phase flow". Mathematics and Computers in Simulation, Vol. 23, (1981).

/8/ Markatos, N.C., Phelps, P.J., Purslow, B.: "Computer simulation of the thermal-hydraulic behaviour of fast reactor pools". Annals of Nuclear Energy, Vol 9, No. 179 (1982).

TABLES

Table 1. Main dimensions of the intermediate heat exchanger.

Outer diameter tubes	21 mm
Wall thickness tubes	1.4 mm
Tube length bewteen tube plates	7.35 mm
Pitch	27 mm
Number of tubes	846
Outer diameter central tube	460 mm
Inner diameter flow shroud (equivalent)	967.6 mm
Inner diameter inner shell	1108 mm
Inner diameter vessel	1329 mm
Outer diameter vessel	1357 mm
Nozzle diameter	300 mm
Material	stainless steel type 304

Table 2. Main dimensions of the steam generator.

Outer diameter tubes	17.2	mm
Wall thickness tubes	2.9	mm
Tube length between tube plates	13.88	mm
Pitch	27.5	mm
Number of tubes	139	
Inner diameter flow shroud	370	mm
Inner diameter vessel	468	mm
Outer diameter vessel	508	mm
Inner diameter inlet and outlet plena	650	mm
Wall thickness plena	25	mm
Diameter sodium nozzles	200	mm
Diameter water/steam nozzles	150	mm
Material	10Cr MoNbNi 9.10 (Nb stabilised 2¼ Cr-1Mo steel)	

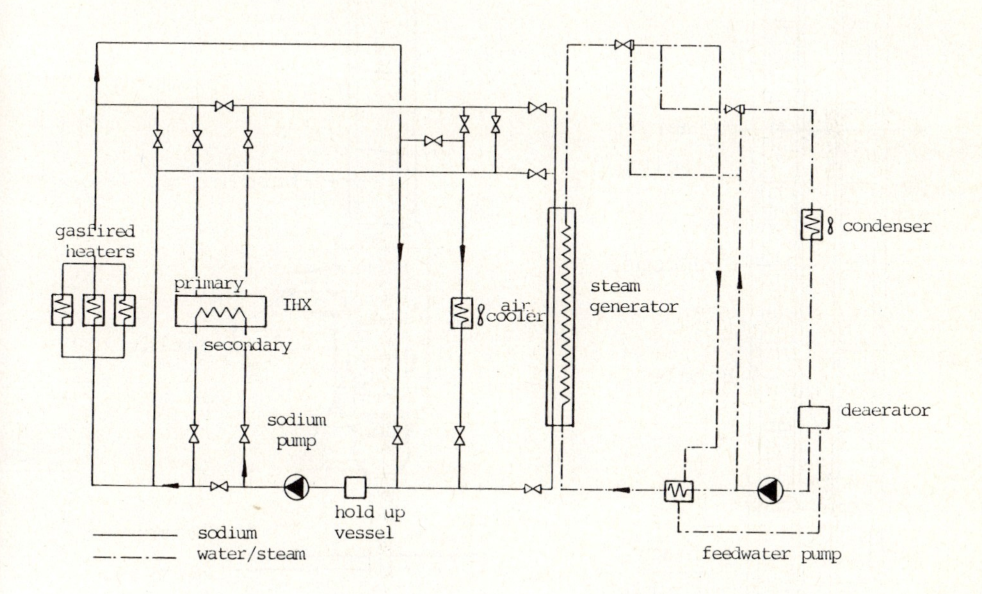

Fig.1 Basic diagram of the 50 MW_{th} testfacility

Fig. 2 Geometry of the prototype SNR-300 IHX

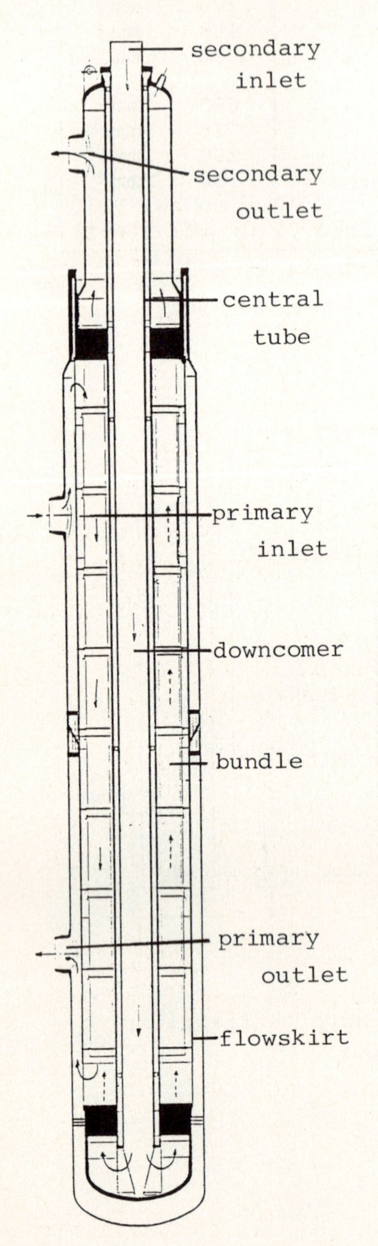

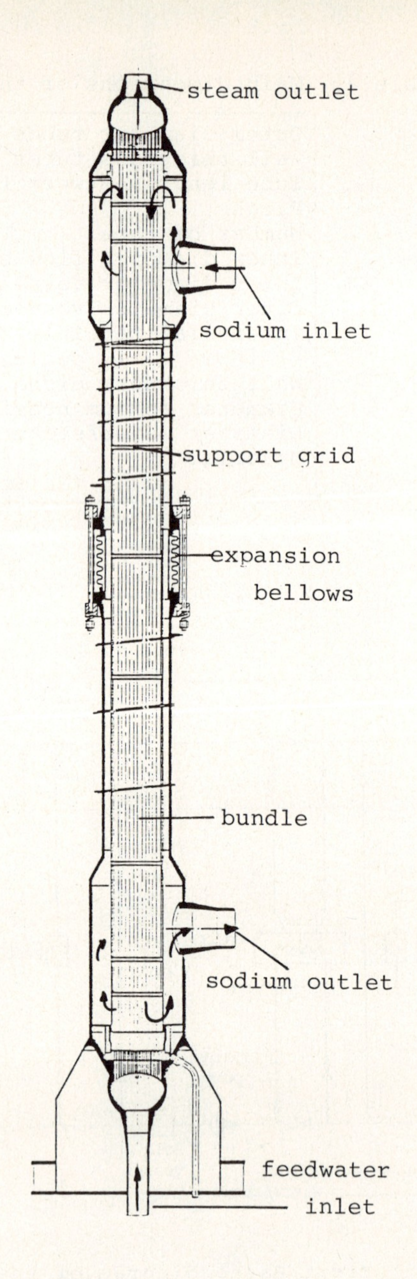

Fig. 3 Geometry of the prototype SNR-300 straight tube superheater

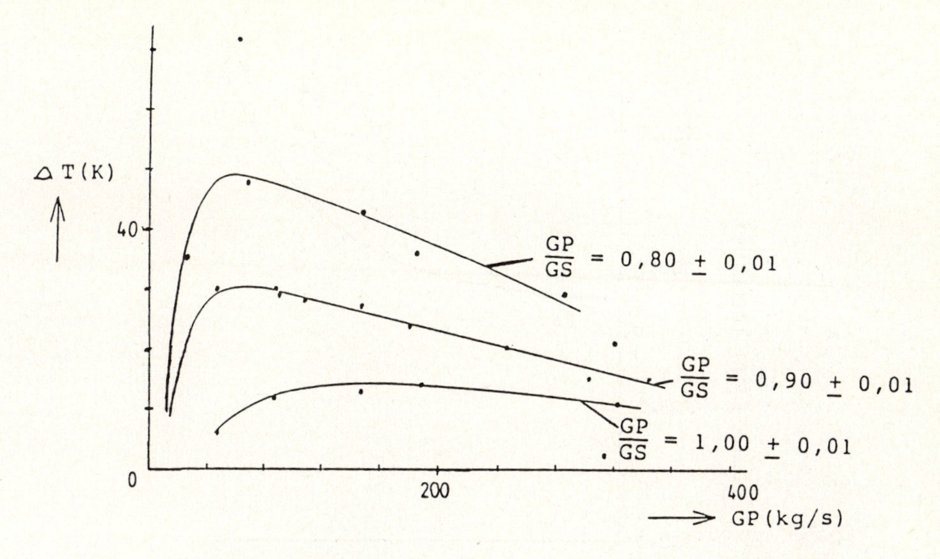

Fig. 4 Temperature differences at the tube outlets as a function of the primary massflow

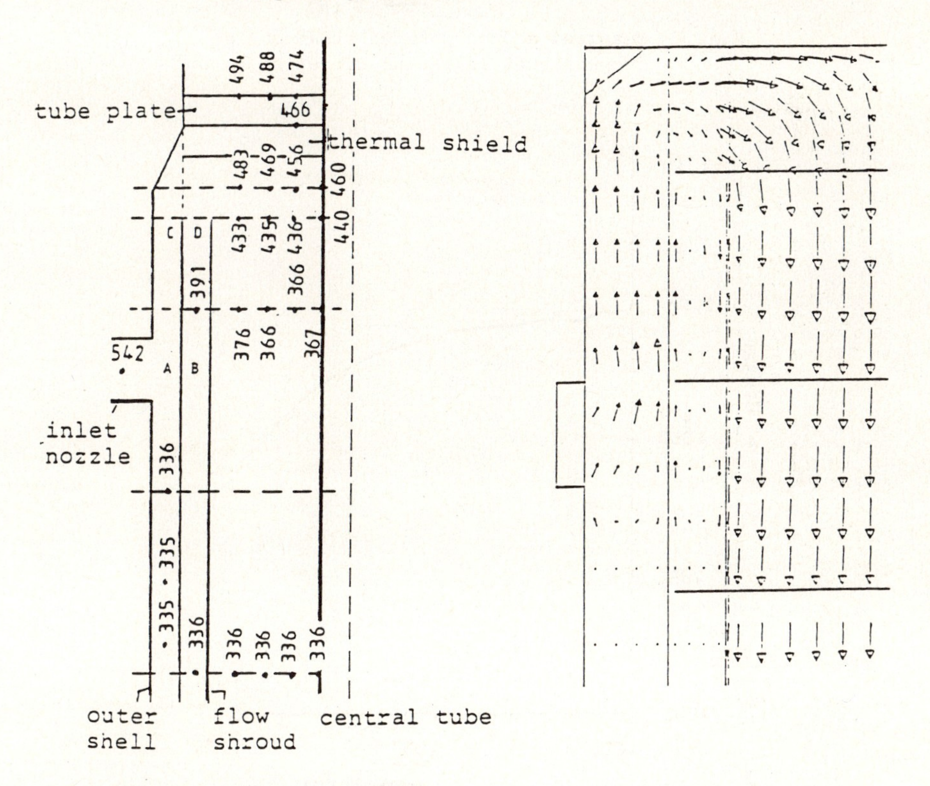

Fig. 5 Measured temperature field (°C) in the top region of the IHX

Fig. 6 Calculated velocity field in the top region of the IHX

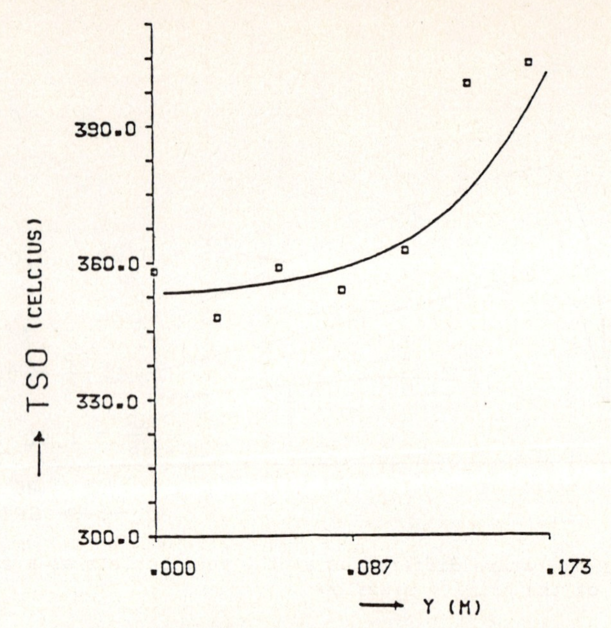

Fig. 7 Measured and calculated steam outlet temperatures in radial direction

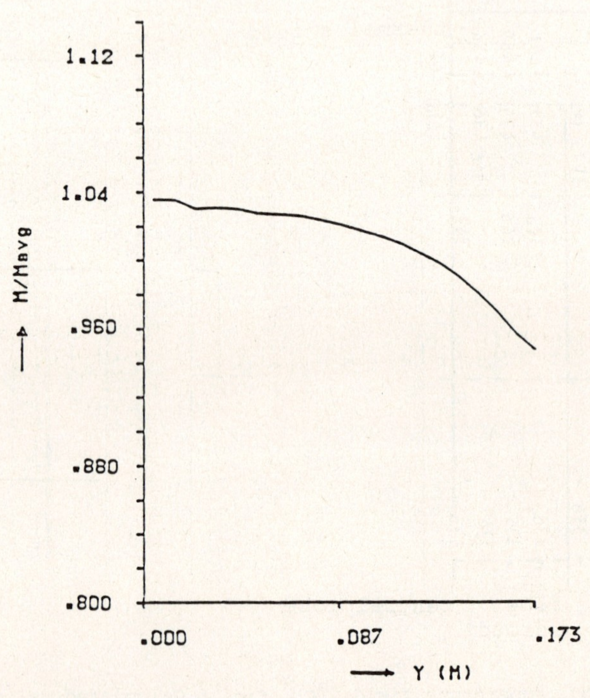

Fig. 8 Calculated feedwaterflow distribution in radial direction

COMPUTATION OF STEADY VISCOUS FLOW NEAR A SHIP'S STERN

M. Hoekstra

Maritime Research Institute Netherlands

P.O. Box 28, 6700 AA Wageningen, The Netherlands

SUMMARY

We consider the problem of computing the steady incompressible viscous flow past the rear part of a ship when free surface effects can be neglected. First some alternative approaches are reviewed which have emanated from different views on how to deal with the pressure in the primitive variable formulation of the Navier-Stokes equations. Then a particular solution method is described. The underlying mathematical model is a slightly reduced form of the Navier-Stokes equations: a main stream direction is identified and diffusion effects in this direction are neglected. The equations are solved in a multiple-sweep space-marching process. Multiple sweeps (global relaxation) are needed to allow the pressure to have influence on the upstream flow field. In each sweep the step-by-step evaluation of the solution is governed by an incomplete factorisation scheme. With this scheme a simultaneous solution is obtained in planes approximately perpendicular to the main stream direction. The performance of the method may be judged from some results of application. The proposed numerical solution to a classical problem in ship hydrodynamics is expected to be of great significance in future ship design studies.

1. INTRODUCTION

Computational fluid dynamics (CFD) is beginning to have some impact on the design of ships. To an outsider, noticing the omnipresence of computers in the modern world, this may seem a modest, if not an insignificant achievement. He should bear in mind, though, that the development of an adequate algorithm for the solution of a given set of partial differential equations - a difficult task in itself - does not warrant an immediate application in a design process. When practically relevant problems have to be solved there are a few more things to be considered. Notably due attention should be paid to the specification and manipulation of geometry, including grid generation, and to the easy visualisation of the results in order to make the tools and/or the results of CFD acceptable. Add to this that the designing of ships is usually 99% conservative and 1% innovative and the statement at the beginning indicates actually something like a break-through.

So far, the potential flow solvers (panel methods) have been the tools most often used in practical design. The improvement of forebody shapes, bossing design for twin-screw ships, the prediction of the ship-generated wave pattern and the design of mooring systems (wave diffraction problems) are typical areas in which they have proved to be useful. For viscous flow predictions boundary-layer methods are now widely available but they have found only incidental application. The main reason is the complete failure of these methods in predicting the flow near the aft end of a ship. Just that part of the flow is of great importance because it constitutes the environment in which the propulsor of the ship is operating.

However, we foresee a frequent use of boundary-layer codes in combination with a tool that will be the subject of the present paper: a Navier-

Stokes solver that can cope with the difficulties which make boundary-layer methods break down near a ship's stern.

In Section 2 we shall formulate our problem and consider some alternative approaches in solving it. Then the particulars of one such approach will be elucidated and results of application will be shown.

2. PROBLEM FORMULATION AND SOME SOLUTION STRATEGIES

We shall study here the water motion around a ship advancing rectilinearly at a constant speed. The speed is high enough to ensure a high Reynolds number (Re), given the small kinematic viscosity of water. On the other hand it is far below the speed of sound; hence the fluid can be considered effectively incompressible. We shall leave the incorporation of free surface effects for the future and consider here so-called double hull flow in which the free surface has become a free-slip symmetry plane. Moreover we assume the flow to be steady but turbulent. In summary, our problem is the computation of the steady, incompressible, turbulent, high Re flow around a body of a more or less streamlined shape in an unbounded flow domain.

The mathematical model to be used is the Reynolds-averaged incompressible form of the Navier-Stokes equations, comprising a vector momentum equation and a scalar mass conservation equation. Even if we restrict ourselves to formulations in terms of the primitive variables (velocity and pressure) and wish the equations to be solved by a finite difference method, quite a few solution procedures appear in the literature which might be used for our problem. Let us classify them according to the choice of the pressure solution scheme which is a crucial matter in incompressible flow computations. Then three main approaches can be distinguished:

a) Methods using a Poisson equation for the pressure [1].
By taking the divergence of the momentum equation and making use of the continuity equation a Poisson equation for the pressure can be derived, which can replace the continuity equation. A successful implicit solution procedure for the resulting set of equations has not yet been found. Either the equations are uncoupled or one or more terms involving the continuity equation are retained in the Poisson equation. In both cases some relaxation scheme, iterating on the pressure until the velocity field is divergence free, must be devised.

b) Artificial compressibility methods [2].
The time dependent form of the equations is adopted and a time derivative of the pressure, multiplied by a small parameter, is added to the continuity equation. From a given set of initial data the flow evolves in time until a steady state is achieved. Upon reaching that state the time derivative of the pressure vanishes and the incompressible flow equations are satisfied. This approach is popular among aerodynamicists because elements of compressible flow solvers can be applied.

c) Boundary-layer compatible methods [3].
Most high Re flows can be identified as shear layer flows, albeit sometimes very complex ones. For thin shear layers a very successful simplified form of the Navier-Stokes equations exists, viz. Prandtl's boundary-layer theory. In that theory one of the momentum equations is uncoupled from the others and fixes the pressure field once the boundary conditions are specified. In the Navier-Stokes equations all four equations are coupled but in high Re flows the pressure may still be governed primarily by the equation expressing momentum conservation in a direction approximately normal to the shear layer. The essential difference between methods of this category and those under a) and b) is

that the pressure field determination does not interfere with the mass conservation property at any moment.

We have chosen to develop a method in the third category because we think it will turn out to be the most efficient approach for our purposes. In the following we shall describe this method. Space limitations do not permit us to go into all the details. Instead we shall highlight some aspects while leaving others undiscussed. Additional information may be found in [4], [5] and [6].

3. DESCRIPTION OF COMPUTATIONAL METHOD

3.1. Domain decomposition and grid generation

Viscous effects become negligibly small outside the boundary layer and the wake. Therefore we choose a zonal approach to our problem and solve the complicated flow equations in only a small part of the flow domain. This computation domain then is internally bounded by the ship's hull; at its inlet boundary, located somewhere halfway the ship's length, information is received from boundary-layer calculations over the front half of the hull; the outlet boundary is located at some distance behind the hull where the pressure can be supposed to have resumed a practically undisturbed level; externally the solution has to be matched to a potential flow solution. In order not to make this matching too difficult we choose the external boundary of the domain well outside the boundary-layer and wake so that the remaining viscous-inviscid interaction is weak. Of course use is made of the (usually double) symmetry properties of the domain, see Fig. 1.

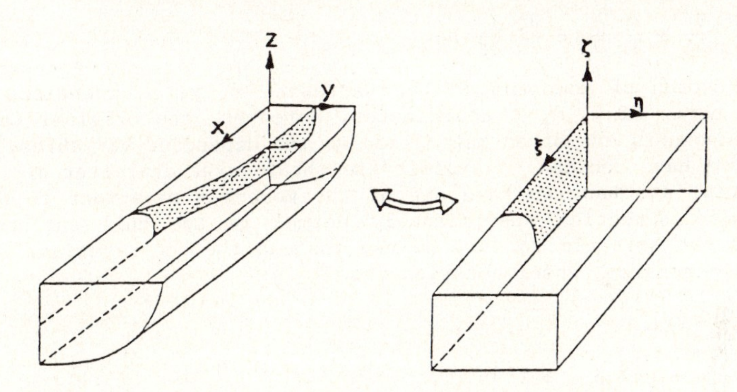

Fig. 1 Computation domain

In this reduced domain a boundary-fitted coordinate system ξ, η, ζ is generated. If we choose $\xi=x$ it suffices to generate grids in a series of transverse cross-sections of the domain which are subsequently connected in longitudinal direction. Non-orthogonality of the grid is implied by the choice $\xi=x$ but in transverse sections an orthogonal-grid generator is applied, i.e. the η, ζ-coordinate lines are mutually perpendicular everywhere except possibly at singular points on the boundaries.

Grid singularities are unavoidable. They are found on knuckle lines of the double hull but also in the wake grid. There we transform a topologically triangular region to a rectangle giving rise to grids as in Fig. 2.

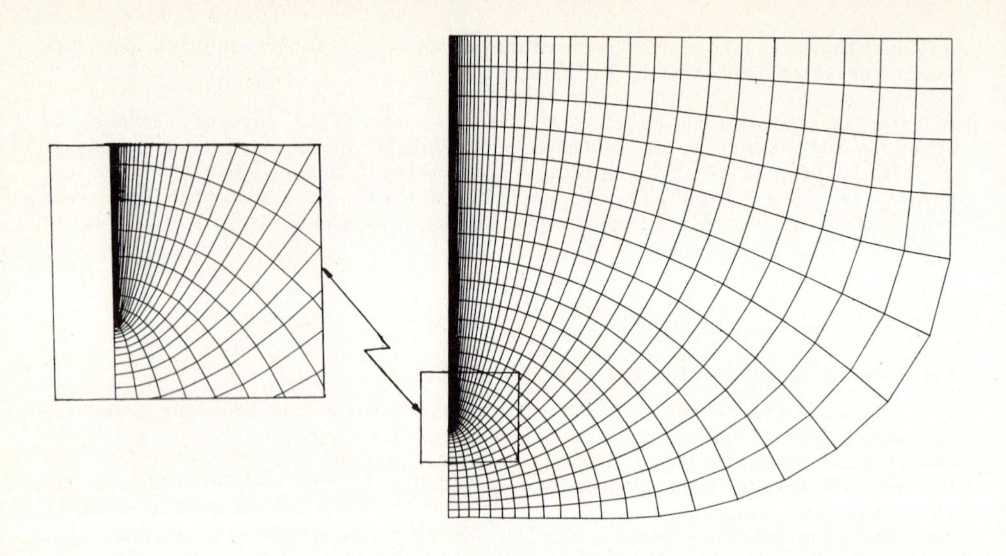

Fig. 2 Grid with singularity in a transverse plane behind the hull

3.2. Equations in curvilinear coordinates

Very often the Navier-Stokes equations in general curvilinear coordinates ξ, η, ζ are obtained by replacing x, y and z derivatives of the Cartesian formulation by the appropriate combinations of ξ, η, ζ derivatives, for example:

$$\frac{\partial}{\partial x} = \frac{\partial \xi}{\partial x}\frac{\partial}{\partial \xi} + \frac{\partial \eta}{\partial x}\frac{\partial}{\partial \eta} + \frac{\partial \zeta}{\partial x}\frac{\partial}{\partial \zeta} \; . \tag{1}$$

Thus conservation of momentum in the Cartesian x, y, z directions is expressed in terms of ξ, η, ζ coordinates. Moreover, the original Cartesian velocity components are often maintained as the dependent variables.

Since we have decided to exploit the shear layer character of the flow to some extent and want to have one of the momentum equations to be associated with a direction approximately normal to the hull surface, this approach is not suitable for us. We use instead the contravariant formulation which in general tensor notation is:

$$\begin{aligned} V^{i}{}_{,i} &= 0 \; . \\ \rho V^{j} V^{i}{}_{,j} &= - g^{ij}\, p_{,j} + \tau^{ij}{}_{,j} \; , \end{aligned} \tag{2}$$

where V^i are the contravariant velocity components, p is the pressure, τ^{ij} is the stress tensor and g^{ij} is the conjugate metric tensor. In expanding these equations to a usable form we choose as the dependent variables the pressure p, the contravariant velocity $u=V^1$ and in the transverse coordinate direction $v=g_{22}V^2$ and $w=g_{33}V^3$. The latter are selected intentionally to obtain regular behaviour near the grid singularity in the wake. Both v and w tend to zero at the singularity even if the vertical (=z) component of the physical velocity is finite there.

The stress tensor contains also the Reynolds stresses which are approximated here by the product of an eddy viscosity and the rate of strain tensor. The eddy viscosity is determined by an analogue of the algebraic

Cebeci and Smith model with adaptations for application in the wake ([7] and [8]).

3.3. Parabolisation

The above system of partial differential equations is elliptic; it cannot be solved as an initial value problem. But it can be shown that the equations become parabolic with respect to the ξ-direction if the following requirements are satisfied:

- the velocity component in ξ-direction is everywhere positive;
- diffusion effects in ξ-direction are omitted;
- the pressure gradient in ξ-direction is known.

The first requirement is usually satisfied in the type of flows that we consider because the ξ-direction is the predominant flow direction. The second requirement can also be satisfied without sacrifycing essential features. It implies dropping some terms which are usually negligibly small. The third requirement is the real stumbling-stone in trying to parabolise the equations for we do not know the pressure field a priori. We can get around this problem by introducing an iteration scheme in which the pressure gradient term is updated successively until convergence is achieved. Thus emerges a multiple-sweep marching solution process: we solve parabolic equations, which allow a downstream marching evaluation of the solution, but by making several sweeps through the domain the elliptic pressure effect is recovered and upon convergence the solution satisfies the original elliptic set of equations.

The exclusion of flow reversal in ξ-direction might seem restrictive but this problem can be alleviated. For if flow reversal does occur it is usually of small magnitude and spatially restricted to a small domain. It can then be included in the flow prediction by deleting locally the relevant convective terms.

3.4. Global pressure relaxation

The question is now: is the proposed iterative process stable and, if so, what is a (rapidly) converging scheme for updating the pressure gradient? This problem has been studied extensively by Rubin and coworkers ([9] and [10]). They have rejected several schemes as being unstable and recommend the use of forward differencing for the discretisation of the pressure gradient in the ξ-momentum equation. The unknown downstream pressure information involved is obtained from a previous sweep or an inital guess. Thus a suitable scheme would be:

$$(p_\xi)_i^n = (p_{i+1}^{n-1} - p_i^n)/\Delta\xi_i \,, \tag{3}$$

if i is a grid point index associated with the ξ-coordinate and n is the iteration count. Unfortunately te scheme is only neutrally stable which can be made plausible by considering the time analogy of the iteration process. Rewriting the above difference formula as

$$(p_{i+1}^{n-1}-p_i^n)/\Delta\xi_i = (p_{i+1}^n-p_i^n)/\Delta\xi_i-(\Delta t/\Delta\xi_i)(p_{i+1}^n-p_{i+1}^{n-1})/\Delta t \,, \tag{4}$$

we can regard it as a discretization of

$$p_\xi - \frac{1}{\alpha} p_t \,, \tag{5}$$

with $\alpha = \Delta\xi_i/\Delta t$. So with the iteration process viewed upon as an evolution process in time, the ξ-momentum equation actually solved is seen to be:

$$L(u) + p_\xi - \frac{1}{\alpha} p_t = 0 \,, \tag{6}$$

where L is a differential operator. The homogeneous equation:

$$p_\xi - \frac{1}{\alpha} p_t = 0 \qquad (7)$$

is recognized as a first order wave equation, allowing pressure waves of arbitrary amplitude to travel undisturbed at a speed α (i.e. $\Delta\xi$ per time step or iteration cycle) in the negative ξ-direction. When the coupling with the other equations is taken into account, the path of the waves in four-dimensional space can be determined but the wave speed is always so as to guarantee a progress of one grid spacing $\Delta\xi$ per time step. These undesirable waves can be eliminated by appropriate boundary conditions for the pressure on those boundaries through which the waves enter the computation domain. Nevertheless it is obvious that for an initial disturbance to run out of the domain we may need as many iterations as there are ξ-stations which is not an attractive perspective.

Israeli and Lin [11] have proposed a remedy which amounts to adding:

$$\sum_{\ell=2}^{i} (p_\ell^n - p_\ell^{n-1})/\Delta\xi_\ell \qquad (8)$$

to the p_ξ term. Notice first that this source term will vanish upon convergence so that it does not disturb the final solution. Considering again the time analogy of the iteration scheme the ξ-momentum equation now reads effectively:

$$L(u) + p_\xi - \int_{\xi_1}^{\xi_{i+1}} \frac{1}{\beta} p_t \, d\xi' = 0 \ , \qquad (9)$$

with $\beta = \Delta\xi_i^2/\Delta t$.

By solving this equation we satisfy locally a discretization of

$$L(u)_\xi + p_{\xi\xi} - \frac{1}{\beta} p_t = 0 \ , \qquad (10)$$

the homogeneous part of which has exponential decay solutions. This suggests a better convergence of the global pressure iteration, which has been confirmed by numerical experiments.

Still pressure influences cannot propagate faster than $\Delta\xi$ per iteration cycle. Therefore the convergence is further enhanced by application of grid sequencing: the first sweeps are made on a grid that is very coarse in ξ-direction; the grid is subsequently refined in a few stages until $\Delta\xi$ is as small as is needed for accuracy.

3.5. Further discretizations and boundary conditions

The discretization of the other terms in the equations may be summarized as follows:

- a non-staggered grid is used, i.e. all variables are defined in the same grid nodes;
- when i, j, k are the grid point indices, the ξ and ζ-momentum equations are centered in i, j, k the η-momentum equation in i, $j+\frac{1}{2}$, k and the continuity equation in i, $j-\frac{1}{2}$, k;
- ξ-derivatives (except p_ξ) are discretized with a three-point backward difference formula;
- two or three-point central differences are used for the η-derivatives;
- ζ-derivatives are represented by central differences or the QUICK scheme [12];
- Newton linearisation is applied to terms involving products of the variables so that the coupling between the equations is preserved in the linearised equations.

Boundary conditions are needed for the velocity components on the inlet plane and for the pressure on the outlet plane. On the external boundary the conditions are boundary-layer like: the pressure and two velocity components are specified. On the hull surface all velocity components are known; the pressure need not be given. Conditions on other boundaries follow from symmetry considerations.

We notice that the hull boundary conditions are indeed applied as indicated; the use of wall functions is deliberately avoided.

The initial pressure field is derived from the boundary conditions assuming $p_\eta = 0$.

3.6. Local iterative solution

Recalling the space-marching character of the solution process, we have to solve at each ξ-station a coupled set of linearized equations, conveniently written in a (4x4) block matrix system as:

$$A\phi = b \tag{11}$$

where A is a sparse matrix, ϕ is the vector of unknowns and b contains known information. In our case A has five non-zero diagonals, associated with the grid nodes (i,j,k), (i,j-1,k), (i,j+1,k), (i,j,k-1) and (i,j,k+1). Some other grid nodes are involved in the discretization as well but the corresponding entries are incorporated in b using previous iterates for ϕ.

There are several iterative algorithms for solving (11). Our method belongs to the class based on the incomplete factorization.

$$A = LD^{-1}U - E \tag{12}$$

where L is a lower triangular, U is an upper triangular and D is a diagonal block matrix. The matrix E signifies the incompleteness or approximate nature of the factorisation. L, U and D are defined by:

- diag (L) = diag (U) = D
- off-diagonal elements of L and U are equal to the corresponding elements of A .
- diag $(LD^{-1}U)$ = diag (A)

The matrix E is bi-diagonal. Given the structure of A:

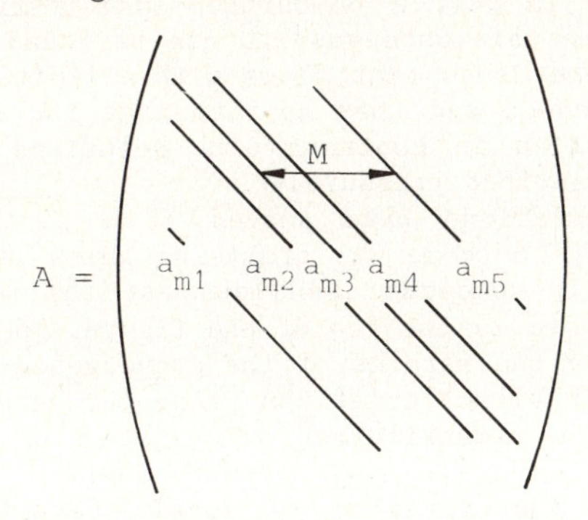

the elements of D can be calculated form the recursion formula:

$$d_m = a_{m,3} - a_{m-1,4}\, a_{m,2}\, d_{m-1}^{-1} - a_{m-M,5}\, a_{m,1} d_{m-M}^{-1} \, . \tag{13}$$

The solution is obtained in two steps:

$$L\psi \quad = b + E\phi \tag{14}$$

$$D^{-1}U\phi = \psi \tag{15}$$

where $E\phi$ is evaluated with data of the previous iteration.

In spite of the fact that E is only bidiagonal the decomposition may not always be well-conditioned. Therefore a stabilization of the pivot matrix D, as proposed by Van der Vorst [13] for scalar matrices, is applied in a block-matrix form.

Iteration is necessary not only because of the incompleteness of the factorisation but also because of the non-linearity of the equations. However, the updating of $LD^{-1}U$ required for the latter only is not carried out in every iteration to reduce the computation time.

A further gain in efficiency is achieved by taking the continuity equation apart and partitioning the original system (11) into:

$$\begin{pmatrix} A_{11} & A_{12} \\ A_{22} & A_{22} \end{pmatrix} \begin{pmatrix} \phi_1 \\ \phi_2 \end{pmatrix} = \begin{pmatrix} b_1 \\ b_2 \end{pmatrix} \tag{16}$$

where each of component of the vector ϕ, contains u,w,p and ϕ_2 contains v. The bidiagonal structure of the scalar matrix A_{22} can be utilized to shift diagonals of A_{12}. The incomplete factorization need then be applied to a modified (3x3) block matrix A_{11} only. We stress that this modification does not abandon the coupling between the continuity and the momentum equations.

The type of procedure described here is also called "strongly implicit" because both η and ζ derivatives are implicitly modelled simultaneously in contrast to for instance the one-dimensional ADI factoring. The strong implicitness of the procedure has been found necessary to let the local iteration process converge also under rather severe conditions.

4. APPLICATIONS

Our method has first been applied to a variety of model problems. For example results for the 2-D flow past the trailing edge of a flat plate have been published in [4]. A comparison with a triple-deck solution is included. In the same reference results can be found for a 2-D flow with a separation bubble. They show that flows with a limited amount of separation can be calculated indeed and they confirm that the solution is regular at the point of separation in contrast with solutions of the boundary-layer equations with a prescribed pressure.

These 2-D calculations have served as a first test of the global relaxation scheme. The outcome was promising which is demonstrated with the example in Fig. 3. It concerns the flow past the rear of an axisymmetric body with a shape shown in the top of the figure. In 14 sweeps the pressure settled itself within the margins of the convergence criterion. This criterion, which holds for all calculations presented here, requires the maximum changes in C_p and the dimensionless velocity to be smaller than 10^{-4} and $5*10^{-4}$ respectively.

The incomplete factorisation (= local iteration) scheme cannot be verified in 2-D or axisymmetric computations because it degenerates to the much simpler and direct solution algorithm for block tridiagonal matrix systems. Therefore we proceeded with studying 3-D flows. The calculation of the flow past the aft part of a slender canoe-type hull form shown in Fig. 4 is such an application. From the results of this study, reported in [5] and [6], we reproduce Fig. 5.

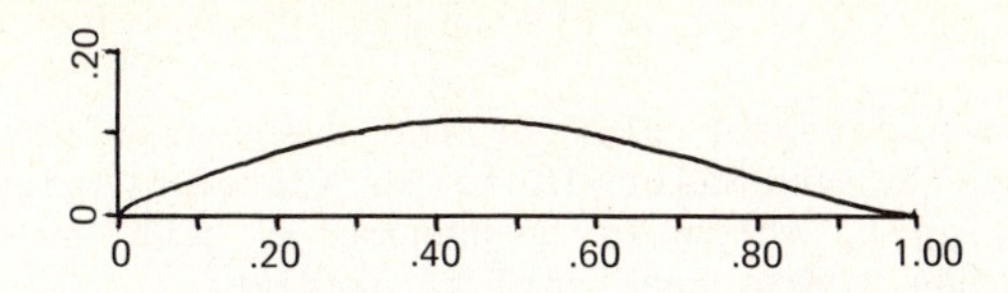

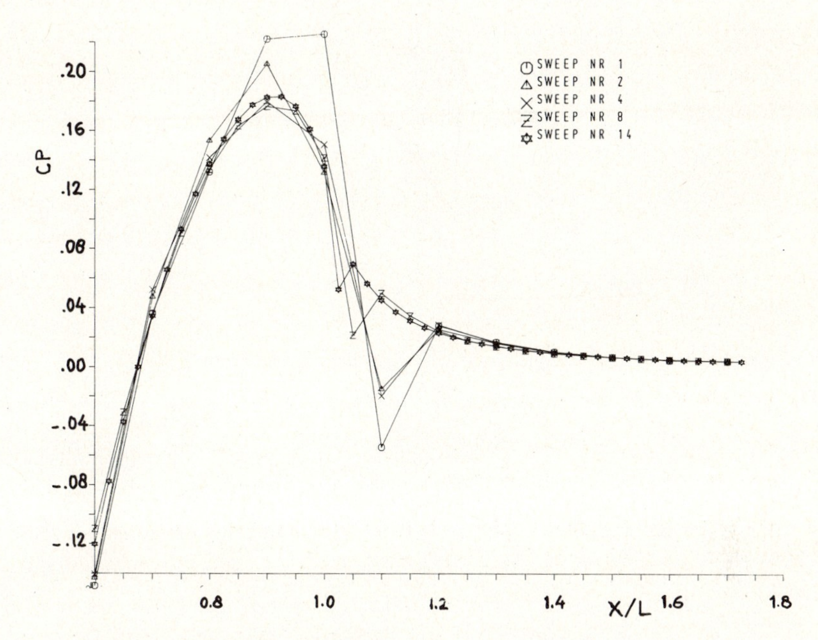

Fig. 3 Convergence of the pressure on the surface and on the wake centre line of an axisymmetric body

It shows the pressure on the hull and on the wake centre line in the summetry plane z=0 as computed in global iteration cycles 2, 5, 8, 9 and 10; the initial guess (sweep No. 0) is also indicated. Again a well-behaved convergence manifests itself.

With regard to the convergence of the local iteration process we can say that it appears to be slower in the wake part than in the forward part of the computation domain. The pivot stabilisation and the particular choice of the velocity variables (see Section 3.2) have, however, improved the behaviour as compared with the results presented in [5]. The convergence rate varies now between 4 and 10 iteration cycles for one order of magnitude reduction of the maximum change in the variables between successive iterations.

Let us finally consider some results of the computation of the flow past a real ship hull. The hull form is represented by a sequence of frame lines in Fig. 6, the forebody to the right and the afterbody to the left. An impression of the finite-difference grid used in this application and comprising 89x40x25 grid nodes can be obtained from Fig. 7. From the results we have selected two plots with information on the velocity field. Fig. 8 shows lines of equal axial velocity in a transverse plane just ahead of the stern. A comparison is made with experimental data which were to the best of our ability derived from boundary-layer measurements. Given the uncertainty in the experimental data, the correspondence is satisfactory and is certainly a lot better than what has been obtained by boundary-layer methods [14].

Fig. 9 shows a vector plot of the transverse velocity components in the plane where the propeller is to be fitted. Obviously a longitudinal vortex has developed, which is a phenomenon well known to occur on many ships but which has always been hard to predict.

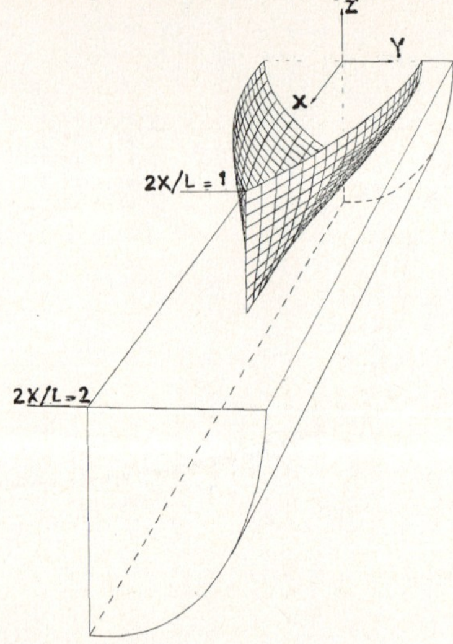

Fig. 4 Aft part of a canoe-type hull form with outline of computation domain

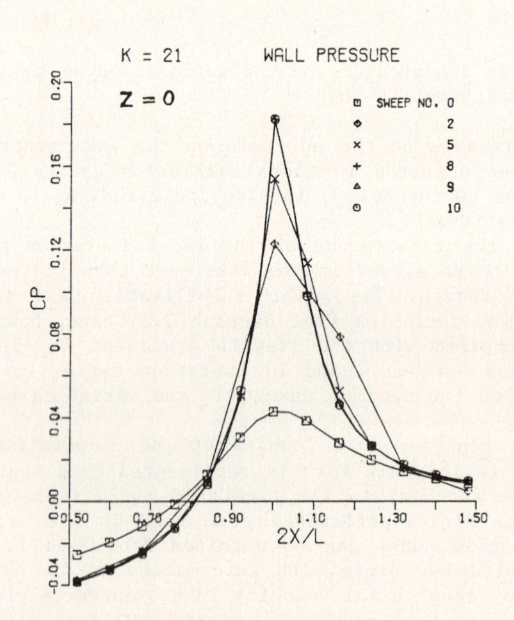

Fig. 5 Convergence of global pressure relaxation process for hull of Fig. 4

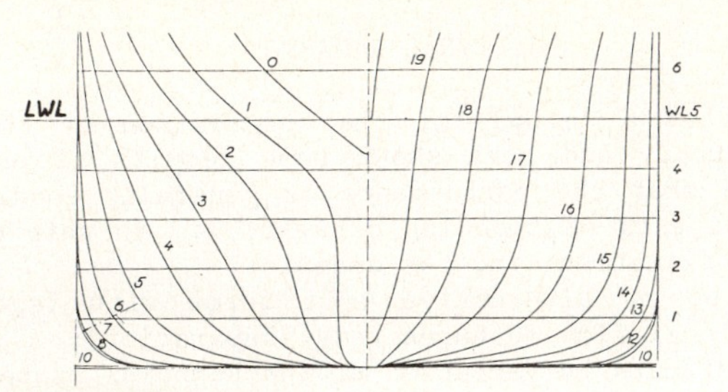

Fig. 6 Body plan of ship's hull

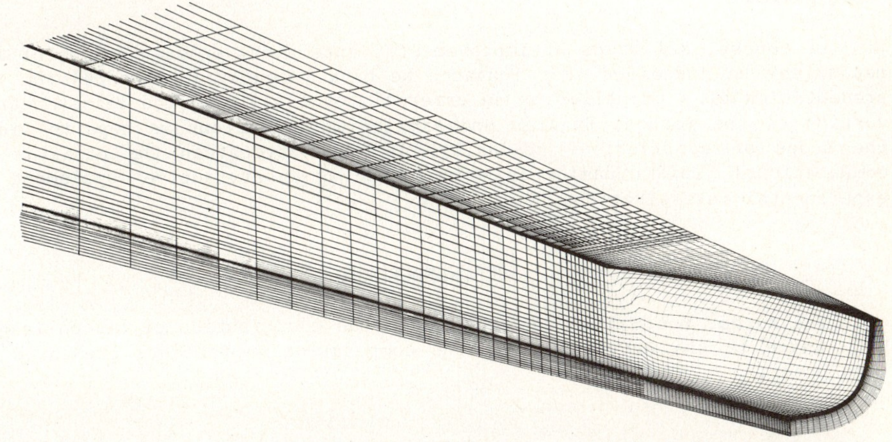

Fig. 7 A view on a part of the grid system for a ship's hull

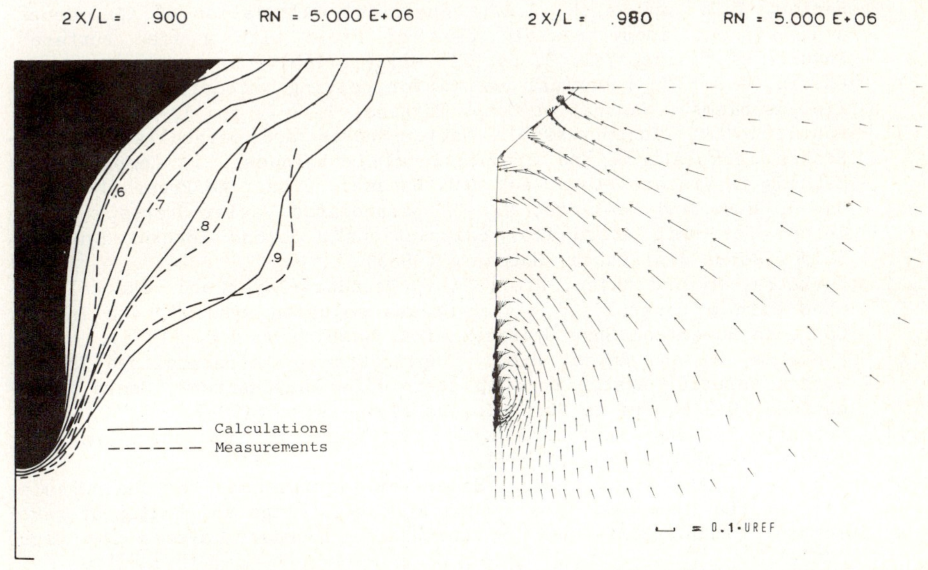

Fig. 8 Lines of equal axial velocity in a transverse plane near the stern of a ship's hull

Fig. 9 Vector plot of transverse velocity components just behind the stern

5. FINAL REMARKS

We have described a method for the calculation of the viscous flow past a ship's afterbody and have shown some results of applications. We cannot but conclude that the performance is generally good. Therefore the method has good prospects of becoming a useful aid in ship design studies. We might think of applications in, for example:
- comparative studies of alternative hull forms with regard to viscous resistance or to local flow features (e.g. separation);
- studies of scale effects by variation of the Reynolds number;
- the calculation of near wake flow data as information for propeller designers.

Of course, all kinds of improvements, extensions or adaptations of the method may be conceived. For instance we have recently implemented the representation of a propeller by an external force field in the axi-symmetric version of the method. By this and similar modifications we hope to widen the scope of applications gradually. It will strengthen the position of computational fluid dynamics in ship design as an attractive supplement to experimental investigations.

ACKNOWLEDGEMENT

The author would like to acknowledge the contribution of his colleague H.C. Raven to the development of the computational procedure presented in this paper.

REFERENCES

[1] Harlow, F.H. and Welch, J.E.: "Numerical calculation of time-dependent viscous incompressible flow of fluid with a free surface", Physics of Fluids, Vol. 8, pp. 2182-2189 (1965).

[2] Chorin, A.J.: "A numerical method for solving incompressible viscous flow problems", Journal of Comp. Physics, Vol. 2, pp. 12-26 (1967).

[3] Rubin, S.G.: "Incompressible Navier-Stokes and parabolised Navier-Stokes formulations and computational techniques", in Computational Methods in Viscous Flows, ed. W.G. Habashi, Pineridge Press (1984).

[4] Raven, H.C. and Hoekstra, M.: "A parabolised Navier-Stokes solution method for ship stern flow calculations", Second Intern. Symp. on Ship Viscous Resistance, Göteborg (1985).

[5] Hoekstra, M. and Raven, H.C.: "Ship boundary-layer and wake calculation with a parabolised Navier-Stokes solution system", 4th Intern. Conf. on Numerical Ship Hydrodynamics, Washington D.C. (1985).

[6] Hoekstra, M. and Raven, H.C.: "Application of a parabolised Navier-Stokes solution system to ship stern flow computation", Osaka International Colloquium on Ship Viscous Flow, Osaka (1985).

[7] Cebeci, T. and Smith, A.M.O.: "Analysis of turbulent boundary layers", Academic Press (1974).

[8] Cebeci, T. and Meier, H.U.: "Modelling requirements for the calculation of the turbulent flow around airfoils, wings and bodies of revolution", AGARD Conference on turbulence boundary-layers, The Hague (1979).

[9] Rubin, S.G. and Lin, A.: "Marching with the parabolised Navier-Stokes equations", Israel Journal of Technology, Vol. 18, pp. 21-31 (1980).

[10] Rubin, S.G. and Reddy, D.R.: "Analysis of global pressure relaxation for flows with strong interaction and separation", Comp. and Fluids, Vol. 11, pp. 281-306 (1983).
[11] Israeli, M. and Lin, A.: "Numerical solution and boundary conditions for boundary layer like flows", 8th International Conference on Num. Methods in Fluid Dynamics, Aachen (1982).
[12] Leonard, B.P.: "A stable and accurate convective modelling procedure based on quadratic upstream interpolation", Comp. Meth. in Appl. Mech. and Eng., Vol. 19, pp. 59-98 (1979).
[13] Vorst, H.A. van der: "Iterative solution methods for certain sparse linear systems with a non-symmetric matrix arising from PDE-problems", Journal of Comp. Physics, Vol. 44, pp. 1-19 (1981).
[14] Larsson, L. ed.: "SSPA-ITTC workshop on ship boundary layers", Publication No. 90, Swedish Maritime Research Centre SSPA (1981).

CONVECTION-DIFFUSION PHENOMENA AND A NAVIER-STOKES PROCESSOR

C.J. Hoogendoorn and Th.H. van der Meer
Technical University Delft, Dept. of Applied Physics
Lorentzweg 1, 2628 CJ Delft,
The Netherlands

SUMMARY

For heat and mass transport in complex flow situations computational methods are very important. Many technological processes can be simulated by a set of convection-diffusion equations. These equations can numerically be solved using a single algorithm based on the finite domain or control volume method. For turbulent transport a k-ε model is often used. This requires that in some cases an experimental validation for a completely new flow situation has to be done.
For two examples the application will be shown. The natural convection in a square cavity both for laminar and turbulent cases will be discussed. For flows and heat transfer in living spaces good predictions including radiative exchange can be given. The second example is the simulation model "Furnace". The flow, combustion and heat transfer in a glass furnace can be predicted. A full 3-dimensional model has been developed. For fine grids, and for time dependent or 3-D situations the computational effort is large. The elliptic flows and the coupling of a large set of partial differential equations give a slow convergence. CPU time on a main-frame computer may run in many hours. This has led us to the development of a processor to directly solve the convection-diffusion algorithm for the finite control volume method as well as the transport equations. This will be applied in a special purpose dedicated Navier-Stokes computer with the capabilities of a super-computer for this special algorithm. It can be expected that this tool will enhance the application of numerical transport phenomena studies strongly.

INTRODUCTION

In heat and mass transfer problems one often is faced with diffusive and convective transport phenomena. This leads to a set of partial differential equations. On the one hand the energy equation to describe heat transport, on the other hand the equations describing the fluid flow. In the case of mass transport of a chemical species in the medium one also has the equation(s) describing the diffusion and convection of that or even more species. Moreover the problem can be time dependent giving transient terms in the equations. For the general case this leads to a set of equations for which it is difficult to obtain solutions by analytical or approximate methods. This is certainly the case for complex geometries.

The increasing need to have a better understanding and analysis of many thermal problems is clear. New technologies, higher qualities of products, require often an accurate prediction of heat transfer. A better thermal design of a process can be important to obtain energy savings as well. This development has led to the application of methods as applied in computational fluid dynamics to convection-diffusion phenomena. Moreover, as will be shown later, the general form of the transport equations is the same for the velocity components to the Navier-Stokes equations as for the temperature (enthalpy) and concentration of a chemical species. This means that the same methods can be applied: i.e. finite differences, finite element and finite domain (or control volume) methods. This latter method as first applied by Spalding [1] and Patankar [2] to heat and mass transfer problems will be discussed in this paper. In two examples the possibilities will be shown. Special attention will be given to the boundary conditions that are typical for heat transport problems. They can complicate the numerical solution procedure. Furthermore the flow in technical applications is turbulent, this requires a modelling of the turbulence for the solution of the flow problem. Turbulence modelling is still in evolution. However a so called k-ε model has obtained wide use for such flows and seems to be the most suitable on the moment. As we are here modelling a physical phenomenon an experimental validation of the solutions obtained can be required. For a number of cases this has been done and led to agreement for the general flow behaviour. A special feature and advantage of the k-ε turbulence modelling is that it leads to the same form of the differential equations for k and ε as mentioned above. A common solution algorithm is possible.
For several practical cases we have found satisfactory solutions for the convection-diffusion equation. For natural convection in cavities Schinkel and Hoogendoorn [3] and Linthorst and Hoogendoorn [4,5] discussed the laminar 2-D and 3-D case. In this paper we discuss the turbulent situation. Bos et. al [6] discussed the turbulent flow of a thermally stratified air layer. A problem with mass transfer in a electro-chemical reactor has been reported by Dalhuysen et. al [7]. Here we will report on a case with mass transfer in a combustion chamber.
Finally it will be discussed that for a general 3 dimensional, transient problem the computational requirements to obtain an accurate solution are very large. Especially if the flow situation has an elliptic character, showing recirculations, convergence of the iterative methods used can be very slow. Also in practical applications the flow and heat and mass transport equations are coupled due to fluid properties like density and viscosity being temperature or concentration dependent. These factors may lead to the need of the use of super-computers to obtain solutions in such cases. The above mentioned complicated cases has led us to develop a special purpose dedicated computer, specifically designed to solve the algorithm as used for the solutions of the equations. We expect that in this way an engineer can obtain for a much lower price and more conveniently solutions for general convective and diffusive heat and mass transfer problems. It has been noted by Singhal [8] that the introduction in industry of computational codes for solution of convective problems is slow and somewhat disappointing. This in contrast to codes for numerical mechanics and in other area's.

He gives several reasons for this slow acceptance by industry of these methods. One of the reasons is undoubtly the large computational effort and high computational costs for a typical complicated problem. We expect that a much less expensive special purpose computer that can be used as a "stand-alone" unit in industry itself may help to overcome this problem. We recently started the development of such a Navier-Stokes processor.

ALGORITHM FOR CONVECTION-DIFFUSION EQUATION

The general form of the convection-diffusion equation for a transient case is:

$$\frac{\partial}{\partial t}(\rho\Phi) + \nabla(\rho\bar{v}\Phi) = \nabla(\Gamma_\Phi \nabla \Phi) + S_\Phi \qquad (1)$$

with respectively the transient, convective, diffusive and source terms for the variable Φ. If Φ represents the velocity components of $\bar{v}$ it leads to the Navier-Stokes equations. With enthalpy and concentration to the energy or mass transport equation. In the equation Γ_Φ gives the diffusivity of the scalar quantity Φ. It stands for laminar or turbulent viscosity in the Navier-Stokes equation. Finally a source term S_Φ can indicate a heat source or sink, a mass source due to a chemical reaction or the pressure gradient and volume forces in the Navier-Stokes equation.

For turbulent flow in general a turbulent diffusive coefficient Γ_Φ has to be used. Only by modelling the turbulent flow predictions of the turbulent transport process are possible. We will use here the k- ε model for turbulent flows.

THE EQUATIONS

In this chapter the equations are defined that simulate a 3-D steady flow with heat and mass transfer. The instantaneous velocity components, the pressure and a scalar variable (representing temperature or concentration) are seperated into a mean and a fluctuating component:

$$\tilde{U}_i = U_i + u_i, \quad \tilde{P} = P + p, \quad \tilde{\Phi} = \Phi + \phi . \qquad (2)$$

The mean components are defined by an averaging over a time long compared to the biggest turbulent time scales of the flow. The in this way averaged equations for continuity, momentum and the scalar variable read (using the Einstein convention):

$$\frac{\partial U_j}{\partial x_j} = 0 \qquad (3)$$

$$\frac{\partial}{\partial x_j}(\rho U_i U_j + \rho\overline{u_i u_j}) + \frac{\partial P}{\partial x_i} = \frac{\partial}{\partial x_j}\left\{\mu\left(\frac{\partial U_i}{\partial x_j} + \frac{\partial U_j}{\partial x_i}\right) + \frac{2}{3}\mu\frac{\partial U_i}{\partial x_j}\delta_{ij}\right\} + g_i\rho \quad (4)$$

$$\frac{\partial}{\partial x_j}(\rho U_j \Phi + \rho\overline{u_j \phi}) = \frac{\partial}{\partial x_j}\left(\Gamma_\Phi \frac{\partial \Phi}{\partial x_j}\right) + S_\Phi \cdot \quad (5)$$

The buoyancy force due to gravitation is included in the momentum equations.
Unknown terms in equations (4) and (5) are the Reynolds stresses: $\rho\overline{u_i u_j}$ and the term representing the turbulent transport of heat and mass: $\rho\overline{u_j \phi}$. The analogy between the Reynolds stresses and the viscous stresses is the basis for the Boussinesq hypotheses defining a turbulent viscosity by:

$$-\rho\overline{u_i u_j} = \mu_t\left(\frac{\partial U_i}{\partial x_j} + \frac{\partial U_j}{\partial x_i}\right) - \frac{2}{3}k\delta_{ij} \cdot \quad (6)$$

And analog to this a turbulent Prandtl number defined by:

$$-\rho\overline{u_j \phi} = \frac{\mu_t}{Pr_{\Phi,t}} \frac{\partial \Phi}{\partial x_j} \cdot \quad (7)$$

The most widely used turbulence model for the determination of the turbulent viscosity is the k-ε model of turbulence. The variables k and ε represent indirectly the characteristic velocity- and length scale of the turbulent fluctuations. Their definitions are:
- kinetic energy of turbulent fluctuations:

$$k = \frac{1}{2}\overline{u_j u_j} \quad (8)$$

- dissipation of kinetic energy of turbulent fluctuations:

$$\varepsilon = \frac{\mu}{\rho}\overline{\frac{\partial u_i}{\partial x_j}\frac{\partial u_j}{\partial x_i}} \cdot \quad (9)$$

With these two variables the turbulent viscosity is defined by:

$$\mu_t = C_\mu \rho \frac{k^2}{\varepsilon} \cdot \quad (10)$$

The equations for k and ε read:

$$\rho\frac{\partial U_j k}{\partial x_j} = \frac{\partial}{\partial x_j}\left\{\left(\frac{\mu_t}{Pr_{k,t}} + \mu\right)\frac{\partial k}{\partial x_j}\right\} + P + G - \rho\varepsilon \quad (11)$$

$$\rho \frac{\partial U_j \varepsilon}{\partial x_j} = \frac{\partial}{\partial x_j} \left\{ \left(\frac{\mu_t}{Pr_{\varepsilon,t}} + \mu \right) \frac{\partial \varepsilon}{\partial x_j} \right\} + (c_{1\varepsilon} P - c_{2\varepsilon} \rho \varepsilon + c_{1\varepsilon} c_{3\varepsilon} G) \frac{\varepsilon}{k} \tag{12}$$

with

$$P = \mu_t \left(\frac{\partial U_i}{\partial x_j} + \frac{\partial U_j}{\partial x_i} \right) \frac{\partial U_i}{\partial x_j} \tag{13}$$

$$G = -\rho \beta \, g_j \frac{\mu_t}{Pr_{\Phi,t}} \frac{\partial \Phi}{\partial x_j} . \tag{14}$$

Values of the constants used in this model are: $C_\mu = 0.09$, $Pr_{k,t} = 1.0$, $Pr_{\varepsilon,t} = 1.3$, $C_{1\varepsilon} = 1.44$, $C_{2\varepsilon} = 1.92$ and $C_{3\varepsilon} = \tanh(|U_2|/|U_1|)$ (if U_2 is in the direction of the gravitation force).
It should be noted that in the averaged momentum equation (eq. (4)) fluctuations in density and viscosity are neglected. This is not the case however for the equations for k and ε, where G represents a source term due to the buoyancy force.
Close to a solid wall very steep gradients exist so many grid points have to be used. This can be prevented if empirical wall functions are being used, which connect the conditions at the wall to the dependent variables just outside the viscous layer. If such wall functions are not available for the application under study, modifications of the k-ε model are necessary in order to include viscous effects close to a wall. Low Reynolds number modifications of the k-ε model are available for this reason (see [9]).

THE NUMERICAL SCHEME

The equations described above are solved numerically. The solution domain is divided into a number of control volumes (surrounding as many gridpoints) over which the convection-diffusion equations are integrated. Using the Gauss-theorem this leads to balances for fluxes across the surface area's of these control volumes. The most attractive feature of this finite domain or control volume method is the integral conservation of the variable over each control volume and thus over the complete solution domain. Staggered grids are used for the velocity components. The grid points for the velocity components lie on the surfaces of the control volumes for the scalar variables. The solution of the velocity field is obtained by applying the SIMPLE procedure (see Patankar and Spalding [2]) or depending on the application by similar procedures like SIMPLER and SIMPLEC (see Patankar [10] and Doornmaal and Raithby [11]). All these procedures turn the continuity equation into an equation for the pressure correction.
To determine the values of the fluxes at the surface area's of the control volumes various finite difference schemes can be used. Mostly a hybrid scheme is applied being a mixture of central differencing and upwind differencing. The problem of artificial diffusion due to the use of upwind differencing has to be recognized. It means that in regions of recirculation many grid

lines have to be used.
The ultimate difference equations for each control volume are solved by TDMA (tri diagonal matrix algorithm or the Thomas algorithm). The equations for a dependent variable on a single line are solve exactly supposing the values of this variable on the neighbouring lines and the values of the other dependent variables as known. Iteratively a solution for all variables for the whole field is reached. For most applications underrelaxation has to be used to reach convergence. Conventional underrelaxation factors as well as the method of false time steps can be applied. Convergence criteria can be defined by the sum of the absolute residuals of each variable, on mass and energy balances on horizontal and vertical lines, or on a total energy balance over the whole solution domain. Dependent on the application one or a combination of these convergence criteria is applied.

MODELLING OF NATURAL CONVECTION

One of the most common problems studied in numerical heat transfer is natural convective flows in cavities. Here, due to the gravitational term being temperature dependent through the density, a strong coupling between flow and temperature equations occur. In dimensionless form the equations (3) to (5) lead for this case to the dimensionless numbers: Pr and Ra, with $Ra = g\beta\Delta TL^3/\nu a$ the ratio of buoyancy and viscous forces. Often a Boussinesq assumptions is made, meaning that only the density in the gravitational term is temperature dependent. However for our numerical scheme this is not necessary. For the case of a 2-D cavity with laminar flow of air there has been set up an international comparison problem to evaluate the different codes used in different studies. This is the so-called bench-mark case as given by de Vahl Davies and Jones [12] and [13]. It consists of the flow in a square cavity with vertical hot and cold isothermal walls and side walls adiatatic.
We compared our results at Ra numbers of 10^4, 10^5 and 10^6 with the standard "accurate" solution obtained by de Vahl Davies on a uniform fine (80 x 80) grid. All our cases were calculated with a strongly non-uniform grid. This is necessary at higher Ra numbers as the flow gets a boundary layer character with strong gradients of the variables near the wall. Table 1 shows the relative deviations in a computed variable in % of the value of that variable as given by the bench-mark solution. Given are for different grid sizes the deviations in the stream function in the centre point, the maximum horizontal velocity u on the vertical centre line and its relative vertical position y, the maximum vertical velocity v on the horizontal centre line and its relative horizontal position x, the average Nusselt number $\overline{Nu}$, the maximum and minimum Nu values and their relative vertical position. Clearly the accuracy increases with decreasing grid size. Higher Rayleigh numbers give somewhat less accurate results. The Nu_{min} value is the most sensitive in relative deviation. However we have to bear in mind that Nu_{min} is relatively small, for $Ra=10^6$ only 11 % of $\overline{Nu}$, see also Fig. 1. It is only a local value at the corners of the cavity, where the flow changes

Table 1. Relative deviations of computed variables compared to bench-mark solution.

Ra	Grid	Ψ_{mid}	u_{max}	y	v_{max}	x	$\overline{Nu}$	Nu_{max}	y	Nu_{min}	y
10^4	20 x 20	- 2.7	- 1.3	- 2.7	- 0.2	14.3	1.4	3.2	- 4.9	- 0.2	- 1.1
10^4	30 x 30	- 1.1	- 0.6	- 1.8	0.8	- 4.2	0.6	1.4	4.9	- 0.2	- 0.7
10^4	45 x 45	- 1.4	0.0	0.5	0.1	3.4	0.3	0.6	2.8	- 0.2	- 0.5
10^5	20 x 20	2.5	0.6	2.9	1.6	9.1	1.3	5.1	-11.1	11.1	- 0.6
10^5	30 x 30	1.6	- 1.3	1.2	0.9	4.5	0.6	2.3	-14.9	2.2	- 0.4
10^5	45 x 45	0.2	- 1.7	1.6	0.2	1.5	0.2	0.7	4.9	0.8	- 0.2
10^6	20 x 20	4.8	8.0	5.3	- 2.9	-34.8	- 0.2	2.0	-34.7	52.7	- 0.1
10^6	30 x 30	3.6	3.7	3.6	1.2	-16.1	- 0.2	- 0.2	-15.9	20.0	0.0
10^6	45 x 45	1.8	0.9	4.1	0.3	- 2.4	- 0.6	- 2.9	- 1.9	19.0	- 0.5
10^{6*}	20 x 20	5.0	8.2	5.3	- 3.1	-34.8	- 0.1	2.7	-34.7	53.5	- 0.1
10^{6*}	30 x 30	3.4	4.6	3.6	1.1	-16.1	- 0.3	- 0.4	-15.9	20.4	0.0
10^{6*}	45 x 45	1.6	1.3	4.1	1.4	- 2.4	0.0	- 1.7	- 1.9	9.3	- 0.2

*) Calculated with k-ε model

its direction and crosses over to the other side. Numerical diffusion can effect the results in this small part of the cavity. However this has little consequence for the whole T and v-fields. Moreover the for engineering purposes important average Nusselt number $\overline{Nu}$ is accurate to within 1.4% even for the 20 x 20 grid. For adiabatic side walls, the average Nusselt-number (dimensionless energy flux) should be constant for each vertical plane and equal to the value on the isothermal vertical walls. The maximum percentile deviation from $\overline{Nu}$ can be used to check the accuracy of the numerical solution. In this respect our solutions prove to be even more accurate than the bench-mark solution (Table 2). These excellent results can be attributed to our very strong non-linear grid point distribution with a large number of grid points near the walls. This is exactly the place where the maximum deviation occurs in the bench-mark solution. Interestingly one can note from Table 1 that using for the highest Ra (10^6) the k-ε turbulent code, this predicts the laminar

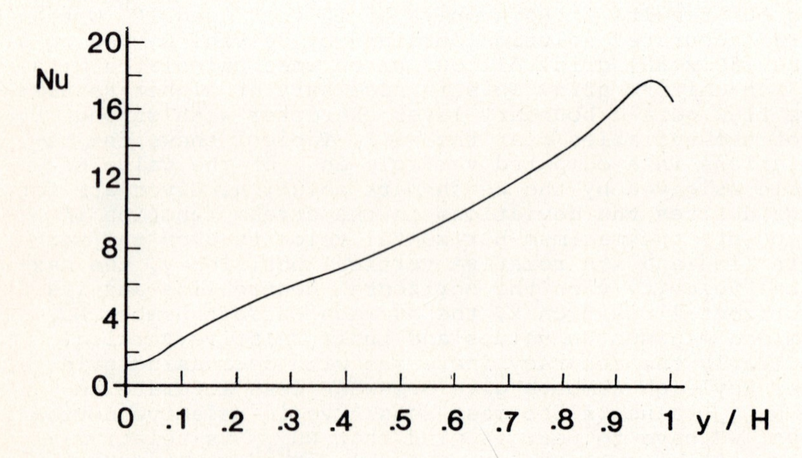

Fig. 1 Nusselt number along cold wall for Ra = 10^6 (45x45)

Table 2. Maximum deviation average Nu-number over vertical gridline.

Ra	benchmark	our results 45 x 45
10^4	.2 %	.009 %
10^5	.2 %	.031 %
10^{6*}	.4 %	.025 %

values equally well as the fully laminar code. Fig.2 shows for Ra= 10^6 the velocity field, giving an arrow for $\underline{v}$ at each grid point. The need for a very fine grid near the isothermal walls is even stronger for turbulent cases, (Ra>10^9).We computed the case of Ra= 3.05 10^{10} with the k- ε model, a 60 x 60 grid has been used. Here it is required that at least one grid point falls between the wall and the velocity maximum of the natural convective boundary layer. The resulting stream line plot (Fig. 3) shows the thin natural convective boundary layer. Along the vertical hot plate the value of the local Nusselt and Rayleigh number, related to the local temperature differences across the boundary layer have been computed. This takes the thermal stratification into account. Fig.4 gives the Nu_x-Ra_x relation obtained, compared to that for a single isothermal vertical plate in an infinite space at constant temperature. Over most of the length x of the wall it agrees well for the two cases, only in the last part the proximity of the side wall effects the local T and v field, compared to the single free plate.
Convergence at high Ra values is relatively slow in this strongly coupled case. For a 2-D case about 1500 iterations are required, each iteration for 10^3 grid points requiring about 1s.

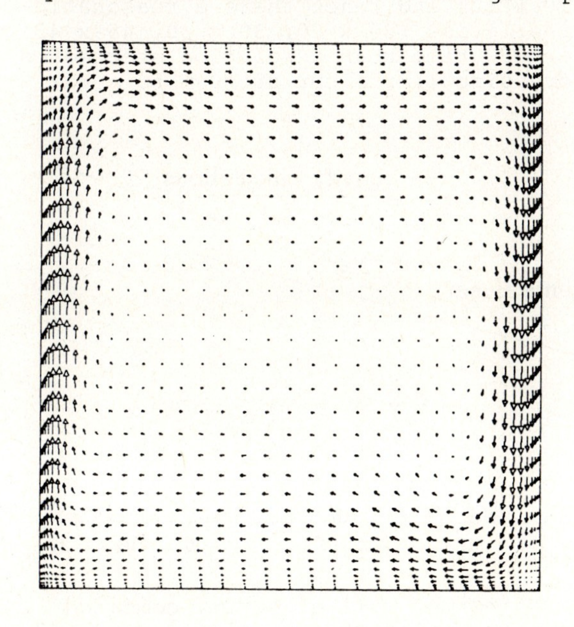

Fig. 2 2-D velocity field for Ra=10^6 from 45 x 45 grid.

Fig. 3 Streamline plot for cavity with Ra= 3.05 10^{10}.

CPU time on a IBM-3083. A 2-D 60 x 60 grid requires about 100 minutes, a 3-D 30 x 30 x 15 grid requires about 2000 iterations and 7 hrs CPU time. A method to improve convergence rate is to start with a coarse grid (10 x 10) and reach there a reasonable convergence. Consecutive finer grids (20 x 20, 30 x 30, 45 x 45) results in a 3 times smaller CPU time than a solution started on a 45 x 45 grid itself. Memory size limits our program to a maxi-

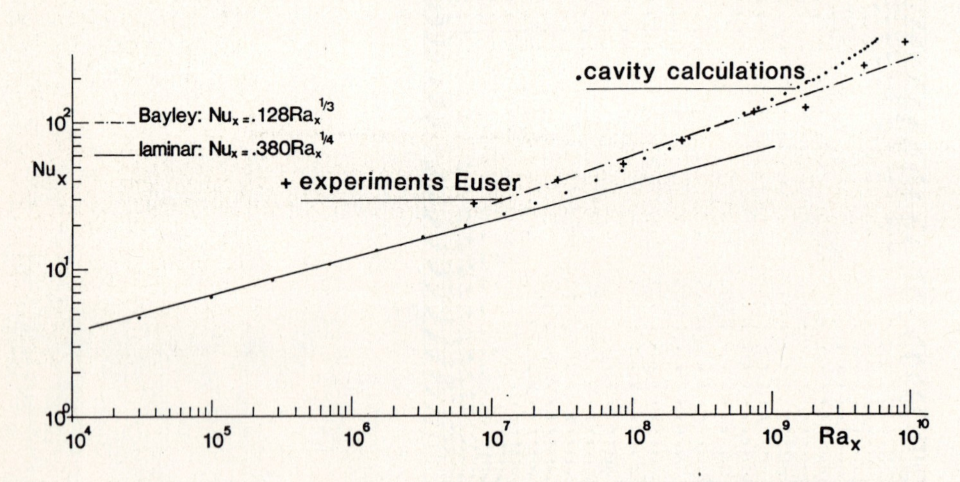

Fig. 4 Nu_x - Ra_x plot for isothermal wall of cavity compared with single isolated vertical plate correlations and experimental data in a climate room by Euser [14]

mum of 40.000 points (45 x 45 x 20 on a 3-D grid).
For practical applications of convective flows in living spaces the numerical simulation for full scale behaviour is very effective. Fig.5a shows the calculated 2-D flow field for a room with a cold window and a hot radiator with a parapet and a window sill. The flow obtained agrees quite well with measured flow fields in a full scale test room. Fig. 5b shows a flow visualization experiment done by ISSO [15] on a practical case with a cold window and a hot radiator below it. In living spaces also radiative heat exchange between wall elements occur, especially for the window and radiator surfaces. This has to be accounted by including radiative exchange between surface elements in the boundary conditions for the convective flow field.

COMBUSTION CHAMBER MODEL FOR GLASS FURNACES

For high temperature heat transfer the combustion of a fossil fuel is often the source for the heating of a process. In the glass melting process, a glass melt is heated from above by a natural gas flame in a combustion chamber (see Fig. 6). Flame development is strongly dependent on the turbulent flow of the gasses in the chamber. Heat transfer is mainly by radiation. Better control of glass quality requires detailed information on the heat transfer and temperatures in the combustion chamber. Actual tests in furnaces through input parameter variation is

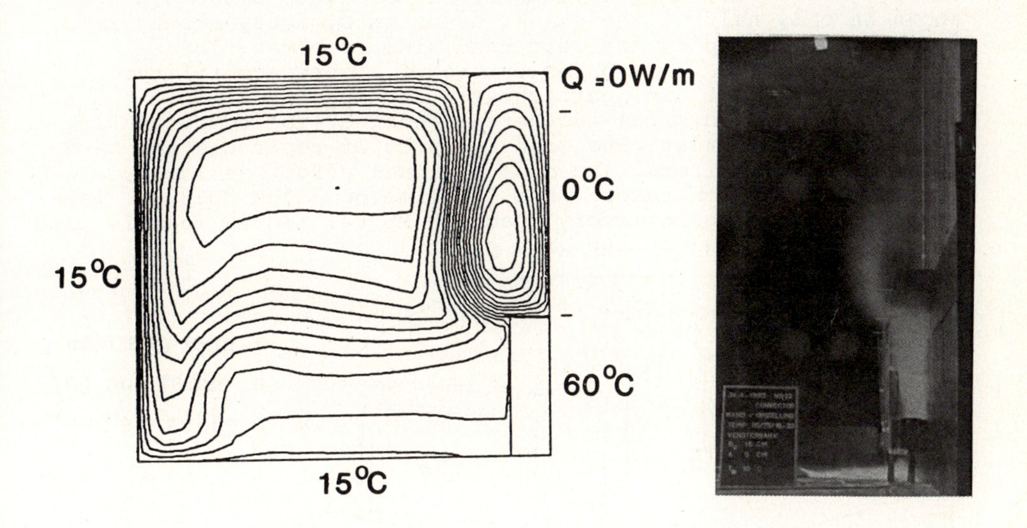

a. b.

Fig. 5a. Computed streamlines (2-D) for a living space with a cold window and a hot radiator below it.
b. Flow visualization experiment for a case similar to that of Fig. 5a. as done by ISSO [15].

impossible in view of the high costs. To make predictions on parameter variation effects we developed a mathematical model FURNACE for the combustion chamber. For the flow and temperature field in the molten glass Simonis [16] developed a fully 3-D laminar simulation. Our aim is ultimately to couple the two models. The combustion chamber model simulates the turbulent 3-D flow field. It is coupled to the energy equation, containing radiative source terms for each volume element. For the flame a two-equation combustion model has been used. This assumes the combustion to be determined by turbulent mixing and by a probability-density function accounting for local air-fuel ratio fluctuations. For radiative transfer a separate zoning model calculates the radiative heat transfer between 36 volume zones and 80 surface zones assuming gray gas approximations. As the radiative fluxes are found from multiple integrals this requires another algorithm than the one for the partial differential equations. With a first assumption of the temperature-field radiative terms are calculated. These are used in the energy equations to find a new T-field and with the coupled flow equations the new flow field. A complete code for this simulation has been set up. Some results for the flow field for cases without combustion are presented here. Figure 7 shows for 2-D case with a 50 x 20 uniform grid a computed flow (streamlines) field, showing recirculation at the top and left hand corner. For a 3-D case Figure 8 gives a profile plot of the main flow (y) direction component of the velocities. A strong forward flow in the flame and a recirculation at the top and at the side has been found. Fig.9 gives a from the 3-D computations obtained flow field in a horizontal cross plane of the chamber. In fact the projection of the 3-D velocity in this (xy) plane has been shown in only half of the plane, assuming symmetry around x= 0.5. Recirculation in the left hand side is clear. These recirculations of flame gases can be of importance for the chemistry of the flame, particularly for NO_x formation. Validation in a physical isothermal model of the computed flow fields is in progress. The main features of recirculation have already been confirmed. The computational effort is large, a 3-D isothermal solution takes 4 hrs CPU-time on a IBM 3083-JX1 main frame computer. Again memory space (5 Mbyte) limits in this case us to the handling of 3000 grid points.

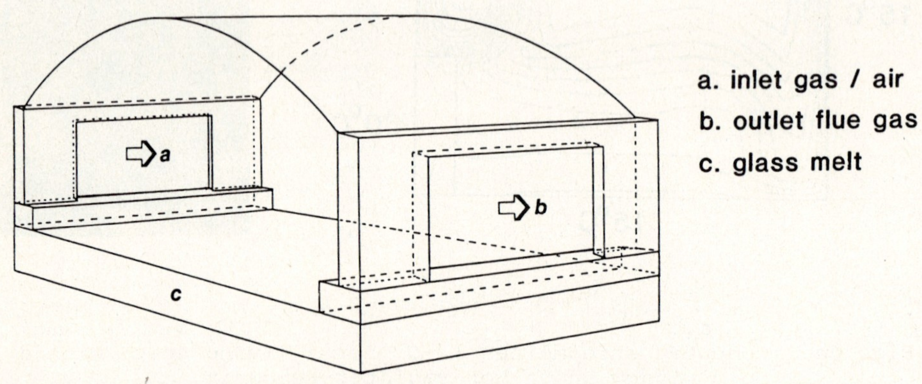

Fig. 6 Perspective view of a combustion chamber of a glass-furnace

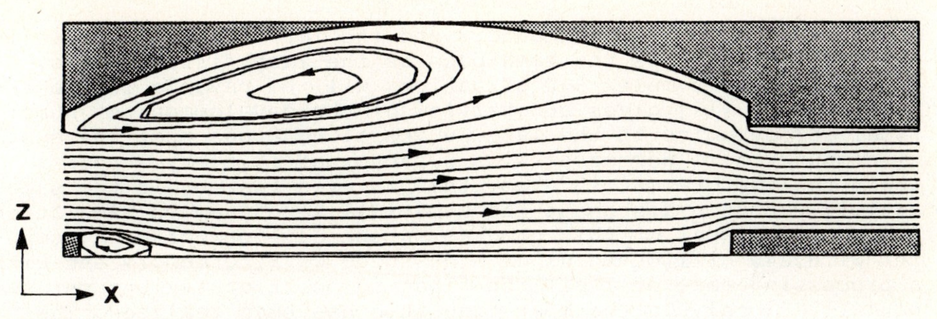

Fig. 7 Two-dimensional streamlines (50 x 50 grid)

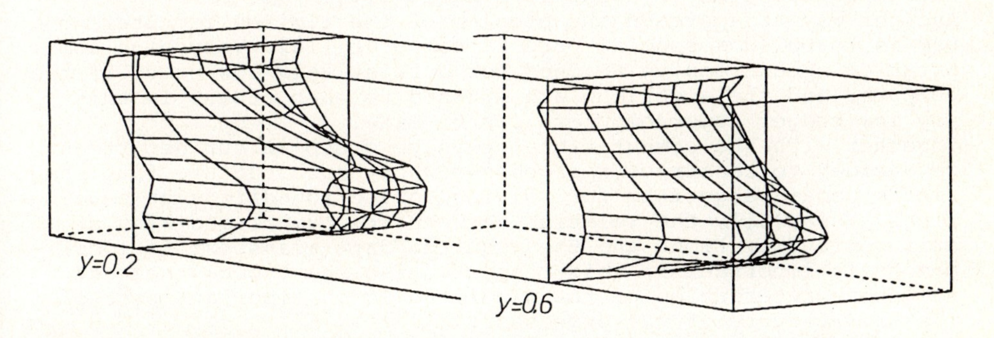

Fig. 8 3-D Profiles of velocity in y-direction at several y-locations (8 x 25 x 10, no plane of symmetry) V_{max} = 3.28 m/s.

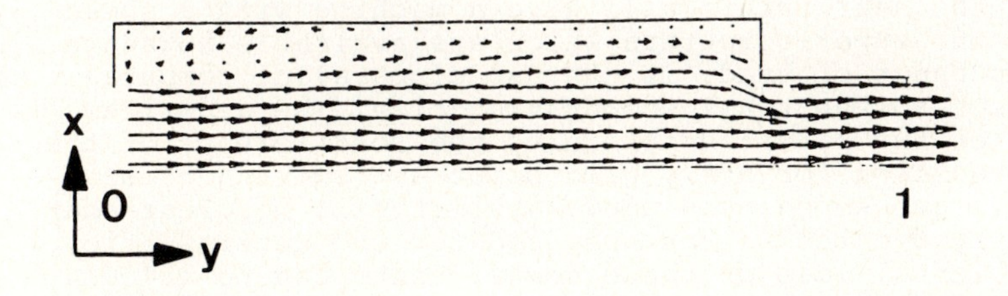

Fig. 9 Projection of 3-D velocity field in xy plane at z = 0.3 from a 10 x 25 x 10 grid (with symmetry plane)

THE NAVIER-STOKES PROCESSOR

From the two examples discussed above, it is clear that much more and much better results can be obtained if more CPU-time is available. Then finer grids can be used. The costs of these calculations however will be high. There are two mechanisms that make better simulations possible in the near future.

If the improvements of computers and the improvements of numerical algorithms keeps the same pace as the last fifteen years, simulations of complex 3-D situations will be payable in due time. Peterson [17] gives an overview of the development of numerical aerodynamics in the last 30 years. He finds that the cost of computer time decreases with a factor of 10 every 7 years, while the improvement in algorithms is even faster. In the same article Peterson discusses the Numerical Aerodynamics Simulation Program. The goal of this program is the development of a computer with 240 million words of high speed working memory and with a processing rate of 1 billion floating point operations per second (1 Gflops). In the mean time this has been realized. The Reynolds-averaged Navier-Stokes equations using 10^7 grid points should be computed in about 10 minutes [17].
Another way to approach the problem of the limited computer power is to build a special purpose computer [18]. Very high computation speeds can be reached for a relatively low price if the computer and the algorithm are attuned to each other. In this way low budget supercomputers can be made.
Together with the computational physics group in our department, we decided to construct a processor whose architecture fits the finite domain algorithm for solving convection-diffusion equations. Especially the relation between the number and kind of arithmetic operations and the required data transport to local memory is important in this respect. Also the typical nearest neighbour interaction of the algorithm is an important part of the architecture of the processor.
Soon one processor card will be come available on which the first of the two examples discussed in this article will be implemented. The maximum computation speed of this processor card can be reached if all adders and multipliers, which are used, continuously produce results. This speed will than be larger than 24 Mflops. Several of the same processor cards parallel to each other in the near future will give a machine with a speed comparable to the super-computers which are available nowadays.
An important subject of study is the use of parallel computations in respect to the total computation time. The algorithm used consists of two parts: determining the coefficients of the differential equation for every grid point and solving these equations. This can be done in many ways. It will be clear that if more recently updated values are used for the determination of the coefficients, less of these coefficients can be calculated parallel. When the Jacobi-algorithm is used, all coefficients and the resulting algebraic equations can be calculated at the same time. If on the other hand a point Gauss-Seidel iteration is used, not all coefficients can be calculated parallel. With a line Gauss-Seidel technique the coefficients on a line can be treated parallel. The resulting tri-diagonal matrix then can be solved in several ways: the conventional TDMA can be used (which means that there is no parallellism), or it can be solved by the method of cyclic reduction [19]. On a sequential computer the second method takes about two times as much CPU-time, but the method is suited for a parallel computer. It is also possible to solve n lines at the same moment. The algorithm deviates from the line Gauss-Seidel algorithm because for blocks of n lines old field values of the last iteration are taken. This can slow down convergence because it takes n iterations before a change in one part of the field affects another part.

For the problem which is being implemented on the processor at this moment it is found that for n=2 convergence is slightly better than for n=1. The convergence rate for n=4 is slightly worse. It is possible that because of a more efficient use of the processor for n=4 the processing time will be the shortest. A second subject of study will be the convergence rate when many grid points are being used. We have already seen that precalculations on coarse grids with interpolations to fine grids can gain a factor of 3 in convergence rate. Calculations on different grids have to be possible on the Navier-Stokes processor and the application of multigrid methods has to be studied. Next to the subjects mentioned much attention will be given to the treatment of the boundary layer along solid walls. In many cases the universal wall functions cannot be used. Since it is possible to do calculations with a large number of gridpoints on the Navier-Stokes processor a better treatment of the flow along walls will be possible.

ACKNOWLEDGEMENT

The above studies have been done with a team of people: A.M. Lankhorst, L. Post and F.F. van der Vlugt. The support by the Foundation for Chemical Research (SON) and for Fundamental Research of Matter (FOM) with financial aid of the Netherlands Organisations for the Advancement of Pure Research (ZWO) and of Technical Research (STW) is appreciated. We acknowledge A.F. Bakker, D.A. v. Delft and B.P.Th. Veltman from the computational physics group for their preliminary discussions about the special purpose processor.

REFERENCES

[1] Spalding, D.B.: "A unified theory of friction, heat transfer in the turbulent boundary layer", Int.J. Heat and Mass Transfer, 7 (1964) pp. 743-761.

[2] Patankar, S.V., Spalding, D.B.: "A calculation procedure for heat, mass and momentum transfer in three dimensional parabolic flows", Int.J. Heat and Mass Transfer, 15 (1972) pp. 1787-1806.

[3] Schinkel, W.M.M., Hoogendoorn, C.J.: "Core stratification effects in inclined cavities", Appl.Sc.Res., 42 (1985) pp. 109-130.

[4] Linthorst, S.J.M., Hoogendoorn, C.J.: "Numerical calculations of heat transfer by natural convection in a cubical enclosure", Proc. 2nd Int. Conf. Num. Methods in Laminar and Turbulent Flow, Venice (1981) pp. 1069-1078.

[5] Linthorst, S.J.M., Hoogendoorn, C.J.: "Natural convective heat transfer in three dimensional inclined small aspect ratio enclosures", Proc. 8th Int. Heat Tr. Conf., San Francisco (1986) pp. 1501-1505.

[6] Bos, W.G., Elsen, T. van den, Hoogendoorn, C.J., Test, F.L.: "Numerical study of a smoke layer in a corridor", Comb. Sc. and Techn., 38 (1984) pp. 227-243.

[7] Dalhuijsen, A.J., Meer, Th.H. van der, Hoogendoorn, C.J., Hoogvliet, J., Bennekom, W.P. van: "Hydrodynamic properties and mass transfer characteristics of electrochemical flow-through cells of the confined wall-jet type", J. Electroanal. Chem., 182 (1985) pp. 295-313.

[8] Singhal, A.K.: "A critical look at the progress in numerical heat transfer and some suggestions for improvement", Num. Heat Transfer, 8 (1985) pp. 505-517.

[9] Patel, V.C., Rodi, W., Scheuerer, G.: "Evaluation of turbulence models for near wall and low-Reynolds number flows", Proc.3rd Symp. Turbulent Shear Flows, Davis, Cal. (1981) pp. 1.1-1.8.

[10] Patankar, S.V.: "Numerical heat transfer and fluid flow, Mc. Graw-Hill, New York (1980).

[11] Doornmaal, J.P. van, Raithby, G.D.: "Enhancement of the SIMPLE method for predicting incompressible fluid flows", Num. heat transfer, 7 (1984) pp. 147-163.

[12] Vahl Davis, G. de, Jones, I.P.: "Natural convection in a square cavity, a comparison excercise", Int.J. Num. Meth. in Fluids, 3 (1983) pp. 227-248.

[13] Jones, I.P.: "A comparison problem for numerical methods in fluid dynamics, the double glazing problem", Num. Meth. in Therm. Probl., Pineridge Press, UK (1979) pp. 338-348.

[14] Euser, H., Hoogendoorn, C.J., Ooyen, H. van: "Airflow in a room as induced by natural convection streams", Energy Cons. in Heating, Cooling, Ventil. Build., Ed. Hoogendoorn, C.J., Afgan, N.H., Hemisphere Publ., USA, 1 (1978) pp. 259-270.

[15] Vermogen van radiatoren bij niet genormeerde opstelling. Publ. nr.1 Stichting ISSO, Rotterdam, Holland (1980) pp. 29-31.

[16] Simonis, F., Waal, H. de, Beerkens, R.C.G.: "Influence of furnace design and operation parameters on the residence time distribution of glass tanks, predicted by 3-D computer simulations", Proc. 14th Int. Conf. on Glass, India (1986).

[17] Peterson, V.L.: "Impact of computers on aerodynamics research and development", Proc. I.E.E.E. 72 (1984) no.1 pp. 68-79.

[18] Hilhorst, H.J., Bakker, A.F., Bruin, C., Compagner, A., Hoogland, A.: "Special purpose computers in physics", J. of Statistical Physics 34 (1984) pp. 987-1000.

[19] Heller, D.: "Some aspects of the cyclic reduction algorithm for block tri-diagonal linear systems", SIAM J. Numer. Anal., 13 (1976) no.4 pp 484-496.

PROBLEMS, ANALYSIS, AND SOLUTIONS OF THE EQUATIONS FOR VISCOELASTIC FLOW

M.A. Hulsen and J. van der Zanden
Delft University of technology
Rotterdamseweg 145, 2628 AL Delft, The Netherlands

SUMMARY

After summarizing the basic equations, the type of the equations for the upper-convected Maxwell model and for the Jeffreys-type models is derived. It is shown that the corotational Maxwell model changes type which is unacceptable from a physical point of view. The Jeffreys-type models (including the Leonov model) have a drastic different type compared to the Maxwell models and are physically more appealing. Correct boundary conditions are briefly discussed for a linearized upper-convected Maxwell model. The boundary conditions for the Jeffreys models are shown to be equal to the boundary conditions for the Navier-Stokes equations, supplemented by boundary conditions for all the extra stresses at the inflow boundary. Jump conditions are derived for Jeffreys-type models. It is shown that in complex flows with sharp corners discontinuities may arise. Numerical methods are discussed that take into account the special type of the equations.

INTRODUCTION

During the last decade, active research is going on in the field of numerical methods for viscoelastic flow [5]. These methods are normally extensions of the methods used for the Navier-Stokes equations and are mostly not based on the basic properties of the governing equations. Most of the methods have convergence problems at high elasticity. It is possible that some of these problems are related to the basic properties of the equations, which are not well understood.

There are only a few papers that pay attention to the mathematical properties of the equations for viscoelastic flow. However, stimulated by the problems in finding suitable numerical methods, the number of papers is growing more rapidly. Rutkevich and Regirer [12,13,14] are the first to analyse basic properties such as type and discontinuities. Ultman and Denn [16] explain the onset of some anomalous heat and momentum transfer with a change of type of the equations.

Recently, there has been more interest in this field and some new papers were published. Luskin [10] analyses a special linearized form of the upper-convected Maxwell model by transforming the system to a canonical form. V.d. Zanden et al. [19] derive the type of the quasi-linear system of an upper-convected Maxwell model for steady flow. Joseph, Renardy and Saut [8] extend this for a class of Oldroyd models. Yoo, Ahrens and Joseph [18] consider the problem of steady fast flow of a family of Maxwell fluids into a hole. They show that the flow is partitioned into elliptic and hyperbolic regions, similar to gas dynamics. Yoo and Joseph [18] analyse the flow of an upper-convected Maxwell model through a channel with wavy walls. For high speed steady flow, the vorticity in the central region has a hyperbolic character and an elliptic character in the wall region.

In order to achieve a better understanding of the basic properties of the equations, the type, the necessary boundary conditions and possible discontinuities are investigated in this paper.

Some remarks are made about the usefulness of various models. Furthermore computational schemes are discussed and some results are presented.

BASIC EQUATIONS

We only consider incompressible flow without heat effects. The equations we need are the balance of linear momentum

$$\rho\dot{\underline{v}} + \text{grad } p - \text{div } \underline{\underline{t}} = \rho\underline{f}, \tag{1}$$

and the conservation of mass

$$\text{div } \underline{v} = 0, \tag{2}$$

where ρ is the mass density, $\underline{v}$ the velocity, $\underline{f}$ the body forces per unit mass, p the hydrostatic pressure and $\underline{\underline{t}}$ is the stress tensor caused by the flow, the so-called extra-stress tensor. A dot above a symbol denotes the material derivative, defined by:

$$(\dot{\ }) = \frac{D}{Dt}(\ ) = \frac{\partial(\ )}{\partial t} + \underline{v}.\text{grad}(\ ) . \tag{3}$$

The balance of angular momentum requires that the extra-stress tensor $\underline{\underline{t}}$ is symmetric if intrinsic moments are absent. Equations (1) and (2) have to be supplemented by a constitutive equation for the extra stress. In the literature many different models for a great variety of fluids can be found.

For large time scales (low frequencies, low deformation rates) both the so-called Maxwell and Jeffreys model have a viscous behaviour. For short time scales (high frequencies, high deformation rates) the Jeffreys model shows a viscous behaviour as well. However, the Maxwell model shows an elastic behaviour for short time scales. The difference in behaviour for short time scales is the fundamental difference between the Maxwell and Jeffreys model. Combining several Maxwell elements in a parallel fashion, yields the so-called generalized Maxwell model. In a similar way, the generalized Jeffreys model can be constructed.

Many Maxwell-type models have been proposed in the literature. For simplicity, we restrict ourselves to the upper-convected Maxwell (UCM) model. This model is given by the following equation:

$$\lambda \frac{\hat{\delta} \underline{\underline{t}}}{\delta t} + \underline{\underline{t}} = 2\eta \underline{\underline{d}} \ , \tag{4}$$

where $\underline{\underline{d}}$ is the symmetric part of the velocity gradient $\underline{\underline{h}}^T = \text{grad } \underline{v}$, λ is a relaxation time ($\lambda \geq 0$), η is a viscosity ($\eta \geq 0$) and the time derivative is given by

$$\frac{\hat{\delta} \underline{\underline{t}}}{\delta t} = \dot{\underline{\underline{t}}} - \underline{\underline{h}}.\underline{\underline{t}} - \underline{\underline{t}}.\underline{\underline{h}}^T. \tag{5}$$

For later use, we rewrite equation (4) in the following way:

$$\lambda \dot{\underline{\underline{t}}} - \underline{\underline{h}}.\underline{\underline{a}} - \underline{\underline{a}}.\underline{\underline{h}}^T + \underline{\underline{t}} = \underline{\underline{0}} \ , \ \underline{\underline{a}} = \underline{\underline{a}}^T = \eta \underline{\underline{1}} + \lambda \underline{\underline{t}}. \tag{6}$$

From the integrated form of the UCM model (Astarita & Marucci 1974), we can easily derive that

$$\underline{\underline{a}} = \frac{\eta}{\lambda} \int_{s=-\infty}^{t} e^{-(t-s)/\lambda} \underline{\underline{C}}^{-1}(s) \, ds \ , \ \lambda \neq 0 \ , \tag{7}$$

where $\underline{\underline{C}}_t(s)$ is the relative Cauchy-Green deformation of a particle volume at time s relative to the current time t. Since $\underline{\underline{C}}_t$ is positive definite [15], it follows from (7) that for $\eta > 0$, $\underline{\underline{a}}$ is positive definite. A major drawback of the UCM model is a singular behaviour of the elongational viscosity and therefore the application range is very limited.

Jeffreys-type models are found by adding a viscosity term to the extra-stress tensor

$$\underline{\underline{t}} = 2\eta_s \underline{\underline{d}} + \underline{\underline{t}}' \ , \tag{8}$$

where η_s is a viscosity ($\eta_s \geq 0$) and $\underline{\underline{t}}'$ satisfies some equation for a Maxwell-type model. We will show that this addition radically changes the type of the equations (in the sense of the classification of partial differential equations). We should note that the Jeffreys model where $\underline{\underline{t}}'$ satifies (4) also has the singular behaviour for elongational flow, similar to the UCM model.

An example of a generalized Jeffreys-type model is the Leonov model [9]. It is a model with multiple modes and for plane flow it can be written as:

$$\underline{\underline{t}} = 2\eta_s \underline{\underline{d}} + \sum_{k=1}^{n} \mu_k (\underline{\underline{b}}_k - \underline{\underline{1}}), \tag{9}$$

$$\theta_k \frac{\hat{\delta} \underline{\underline{b}}_k}{\delta t} + \frac{1}{2}(\underline{\underline{b}}_k^2 - \underline{\underline{1}}) = \underline{\underline{0}} \ , \ k = 1, \ldots, n \ , \tag{10}$$

with the constraints $\det \underline{\underline{b}}_k = 1$. The tensors $\underline{\underline{b}}_k$ are internal deformation tensors and have to be positive definite. The linear model underlying the Leonov model is a generalized Jeffreys model with n modes. The introduction of the term $2\eta_s \underline{\underline{d}}$ is most obvious for modelling dilute polymer solutions. In that case, η_s accounts for the viscosity of the Newtonian solvent. However, a polymeric melt may show a term like $2\eta_s \underline{\underline{d}}$ as well, accounting for the shear between the polymer chains. Moreover, for polymer melts, equation (8) can also be useful for some limited range of shearing rates to take into account small relaxation times. Polymer liquids consist of networks of molecular chains with a large range of length scales and therefore relaxation times. This is the basis of multiple mode models like the Leonov model in (9) - (10). Now assume that some polymer can be described by m relaxation times

$$\theta_1 \geq \theta_2 \geq \theta_3 \cdots \theta_n \geq \theta_{n+1} \cdots \geq \theta_m, \tag{11}$$

with $\eta_s = 0$. For some kind of flow the characteristic time is T and assume that

$$\frac{\theta_{n+1}}{T} \ll 1 . \tag{12}$$

This means that the modes n+1,...,m behave Newtonian and for this type of flow the contribution to the total stress of these modes, can be approximated by a single term $2\eta_s d$ with $\eta_s = \sum_{k=n+1}^{m} \eta_k$. Depending on the number of modes m - n, the value of η_s can be very small with respect to the shear viscosity for zero strain rate $\eta_o = \sum_{k=1}^{m} \eta_k$.
In this section we have only mentioned a few models and for extensive discussion and derivation of these and other models, we refer to books on this subject [2,3,7].

TYPE OF THE SYSTEM OF EQUATIONS FOR SEVERAL MODELS

In order to develop acceptable numerical schemes, it is necessary to gain a good understanding of the mathematical properties of the basic equations. An important property of a system of partial differential equations is whether its characteristics are real or not. The behaviour of the solution is very much dependent on this so-called type. The discussion of the type of Maxwell models can also be found in a somewhat different form, in the work of v.d. Zanden et al. [19] and Joseph et al. [8]. We will also discuss the Jeffreys models, which appear to be of a simpler type, being physically and numerically more appealing.

For simplicity, we restrict ourselves to plane flow (for more details see the report by Hulsen [20]). For the time-dependent case, the number of independent variables is three, namely x, y and the time t. The unknowns for the UCM model are u, v, p, t_{xx}, t_{yy}, t_{xy} and the system of equations can be found from (1), (2) and (6):

$$\begin{aligned}
&\rho \frac{\partial u}{\partial t} + \rho u \frac{\partial u}{\partial x} + \rho v \frac{\partial u}{\partial y} + \frac{\partial p}{\partial x} - \frac{\partial t_{xx}}{\partial x} - \frac{\partial t_{xy}}{\partial y} = \rho f_x, \\
&\rho \frac{\partial v}{\partial t} + \rho u \frac{\partial v}{\partial x} + \rho v \frac{\partial v}{\partial y} + \frac{\partial p}{\partial y} - \frac{\partial t_{xy}}{\partial x} - \frac{\partial t_{yy}}{\partial y} = \rho f_y, \\
&\frac{\partial u}{\partial x} + \frac{\partial v}{\partial y} = 0, \\
&\lambda \frac{\partial t_{xx}}{\partial t} + \lambda u \frac{\partial t_{xx}}{\partial x} + \lambda v \frac{\partial t_{xx}}{\partial y} - 2a_{xx} \frac{\partial u}{\partial x} - 2a_{xy} \frac{\partial u}{\partial y} + t_{xx} = 0, \\
&\lambda \frac{\partial t_{yy}}{\partial t} + \lambda u \frac{\partial t_{yy}}{\partial x} + \lambda v \frac{\partial t_{yy}}{\partial y} - 2a_{xy} \frac{\partial v}{\partial y} - 2a_{yy} \frac{\partial v}{\partial y} + t_{yy} = 0, \\
&\lambda \frac{\partial t_{xy}}{\partial t} + \lambda u \frac{\partial t_{xy}}{\partial x} + \partial v \frac{\partial t_{xy}}{\partial y} - a_{xx} \frac{\partial v}{\partial x} - a_{yy} \frac{\partial u}{\partial y} + t_{xy} = 0.
\end{aligned} \tag{13}$$

This is a quasi-linear system and its characteristic equation can be found by application of standard procedures [4]. After lengthy calculations the following simple equation can be found for the characteristic surface $\phi(x,y,t) = 0$

$$Q(\phi_i) = (\lambda\xi)^2(\phi_x^2 + \phi_y^2)(a_{xx}\phi_x^2 + a_{yy}\phi_y^2 + 2a_{xy}\phi_x\phi_y - \lambda\rho\xi^2) = 0, \tag{14}$$

where $\xi = \dot{\phi} = \phi_t + u\phi_x + v\phi_y$ and the Cartesian components of the normal vector are denoted by $(\phi_x,\phi_y,\phi_t) = (\partial\phi/\partial x,\partial\phi/\partial y,\partial\phi/\partial t)$. This equation has also been derived in v.d. Zanden et al. [19] and Joseph et al. [8].

The equation (14) is composed of three factors, being treated separately both for the unsteady and the steady cases. In the latter case there are only two independent variables x and y. The corresponding characteristic equation can be found from (14) putting $\phi_t = 0$.

The first factor in (14) gives:

$$\lambda\xi = \lambda(\phi_t + u\phi_x + v\phi_y) = 0 . \tag{15}$$

This means that surfaces consisting of particle trajectories in the x, y, t-space are double characteristic surfaces. In the steady case streamlines are double characteristic curves. It is remarkable that particle trajectories are only double characteristics, since three stress components are determined by the deformation history. The constitutive equation considered separately has three real characteristics which are all particle trajectories. In the next chapter we will see that this is also important for the formulation of the boundary conditions.

The second factor in (14) yields:

$$\phi_x^2 + \phi_y^2 = 0 , \tag{16}$$

which leads to the solution $\phi_x = \phi_y = 0$, $\phi_t \neq 0$. This means that the x,y-plane is a characteristic surface. This is typical for parabolic equations (e.g. the heat equation). For the steady case, equation (16) does not have real solutions.

The most complicated factor of (14) is the last one and gives:

$$a_{xx}\phi_x^2 + a_{yy}\phi_y^2 + 2a_{xy}\phi_x\phi_y - \lambda\rho\xi^2 = 0 . \tag{17}$$

This equation is the characteristic equation for the following prototype equation:

$$a_{xx}\frac{\partial^2 w}{\partial x^2} + a_{yy}\frac{\partial^2 w}{\partial y^2} + 2a_{xy}\frac{\partial^2 w}{\partial x\partial y} = \lambda\rho\frac{D^2 w}{Dt^2} . \tag{18}$$

This is a hyperbolic wave equation with convection and non-isotropic wave speed properties.

We can simplify (17) by transformation to the principal axes of $\underline{\underline{a}}$ (and $\underline{\underline{t}}$) and by the introduction of principal wave speeds:

$$c_1^2\phi_1 + c_2^2\phi_2 - (\phi_t + v_1\phi_1 + v_2\phi_2)^2 = 0 \ , \tag{19}$$

where the principal directions of the tensor $\underline{\underline{a}}/\lambda\rho$ are denoted by the indices 1 and 2, and the corresponding principal values by c_1^2 and c_2^2, hence

$$c_i^2 = a_i/\lambda\rho = (t_i + \eta/\lambda)/\rho, \quad i = 1, 2, \tag{20}$$

a_i and t_i being the principal values of $\underline{\underline{a}}$ and $\underline{\underline{t}}$ respectively. As we have already noted, the tensor $\underline{\underline{a}}$ is a positive definite tensor for the UCM model, hence the wave speeds c_1 and c_2 are real. Real-valued wave speeds are required for a well-posed initial value problem for (18) [4]. Dupret, Marchal and Crochet [6] show that in numerical calculations with so-called mixed methods, the tensor $\underline{\underline{a}}$ can become not positive-definite because of approximation errors. They also show that this numerical "change of type" has a dramatic effect on the flow field. Wiggles are produced and convergence is lost.

For the steady case we get from (19):

$$c_1^2\phi_1 + c_2^2\phi_2 - (v_1\phi_1 + v_2\phi_2)^2 = 0 \ . \tag{21}$$

It is easy to show that only real-valued solutions exist if:

$$\frac{v_1^2}{c_1^2} + \frac{v_2^2}{c_2^2} \geq 1 \ . \tag{22}$$

This is equal to the condition for the tensor $\underline{\underline{a}} - \lambda\rho\, \underline{v}\, \underline{v}$ being not positive definite. If the velocity $\underline{v}$ satisfies the equal-sign in (21), then change of type occurs, similar to the change of type in transonic flows in gas dynamics. However, the physical cause of the wave speeds is different. In gases this is the compressibility and the inertia and in this case it is the elasticity and the inertia. We call the velocity $\underline{v}$ supercritical, if the greater sign in (21) is satisfied and subcritical if $\underline{v}$ does not satisfy (21). Ultman and Denn [16] and Yoo, Ahrens and Joseph [17] suggest that some experimentally observed phenomena can be explained by this change of type of the PDE.

In order to indicate when the type of the PDE changes, we consider the case $\underline{\underline{t}} = \underline{\underline{0}}$. In that case $c^2 = c_1^2 = c_2^2 = \eta/\lambda\rho$. The inequality (21) can now be written as:

$$Ma^2 = \frac{v^2}{c^2} = ReDe \geq 1, \tag{23}$$

where Ma is the Mach number, $v^2 = v_1^2 + v_2^2$, $Re = \frac{\rho v L}{\eta}$ is the Reynolds number, $De = \frac{\lambda V}{L}$ is the Deborah number and L is a characteristic length scale of the flow. For polymer melts the Reynolds number Re is very low in practical problems: 10^{-5}-10^{-2}. To satisfy (23) the Deborah number De has to be very high. For polymer solutions Re can be much higher, hence satisfying (23) for much lower Deborah numbers.

Summarizing the results of this section, we can say that the total system of equations for the UCM-model is of a parabolic-hyperbolic type in the time-dependent case and of an elliptic-hyperbolic type in the steady case. In the latter case, change of type can occur for certain critical velocities.

In [20] the so-called corotational Maxwell (CRM) model is also analysed. For this model the following characteristic equation can be found:

$$Q(\phi_i) = (\lambda\xi)^2(\phi_x^2 + \phi_y^2)(k_{xx}\phi_x^2 + k_{yy}\phi_y^2 +$$
$$+ 2k_{xy}\phi_x\phi_y - \lambda\rho\xi^2) = 0, \tag{24}$$

where $\underline{\underline{k}}$ is a tensor that depends on the local value of the extra-stress $\underline{\underline{t}}$. The result seems to be completely analogous to the UCM model. However, the tensor $\underline{\underline{k}}$ can become not positive definite and one of the wave speeds becomes imaginary. In that case the wave equation (18) with $\underline{\underline{a}}$ replaced by $\underline{\underline{k}}$ is no longer well-posed as an initial value problem [4] and Hadamard instabilities exist. This is a very strong instability with an infinite number of instability modes, having arbitrarily short wavelengths. The growth rates tend to infinity as the wavelengths tend to zero.

Problems that are not well-posed, are normally rejected as physical models, as has been pointed out by Courant & Hilbert [4]. However, the same authors give examples of meaningful but not well-posed problems. Ahrens, Joseph, Renardy and Renardy [1] try to explain certain abrupt changes in polymer flow (e.g. melt fracture) with instabilities of Hadamard type. With very little experimental evidence so far, it seems wise not to use the CRM model or related models showing this type of instability.

Jeffreys-type models are given by equation (8).Due to the addition of an extra viscosity term the type of the system of equations radically changes, although the magnitude of the extra term can be small. This is comparable with convection dominated flow where the type of the equations is governed by the diffusion terms but the overall behaviour by the convection.

If we take the upper-convected model for $\underline{\underline{t}}'$ in (6) then we get the following system of equations (primes omitted):

$$\rho\frac{\partial u}{\partial t} + \rho u\,\frac{\partial u}{\partial x} + \rho v\,\frac{\partial u}{\partial y} + \frac{\partial p}{\partial x} - \frac{\partial t_{xx}}{\partial x} - \frac{\partial t_{xy}}{\partial y} - \eta_s\,\frac{\partial^2 u}{\partial x^2} - \eta_s\,\frac{\partial^2 u}{\partial y^2} = \rho f_x,$$
$$\rho\frac{\partial v}{\partial t} + \rho u\,\frac{\partial v}{\partial x} + \rho v\,\frac{\partial v}{\partial y} + \frac{\partial p}{\partial y} - \frac{\partial t_{xy}}{\partial x} - \frac{\partial t_{yy}}{\partial y} - \eta_s\,\frac{\partial^2 v}{\partial x^2} - \eta_s\,\frac{\partial^2 v}{\partial y^2} = \rho f_y,$$
$$\frac{\partial u}{\partial x} + \frac{\partial v}{\partial y} = 0,$$
$$\lambda\,\frac{\partial t_{xx}}{\partial t} + \lambda u\,\frac{\partial t_{xx}}{\partial x} + \lambda v\,\frac{\partial t_{xx}}{\partial y} - 2a_{xx}\,\frac{\partial u}{\partial x} - 2a_{xy}\,\frac{\partial u}{\partial y} + t_{xx} = 0, \tag{25}$$
$$\lambda\,\frac{\partial t_{yy}}{\partial t} + \lambda u\,\frac{\partial t_{yy}}{\partial x} + \lambda v\,\frac{\partial t_{yy}}{\partial y} - 2a_{xy}\,\frac{\partial v}{\partial x} - 2a_{yy}\,\frac{\partial v}{\partial y} + t_{yy} = 0,$$
$$\lambda\,\frac{\partial t_{xy}}{\partial t} + \lambda u\,\frac{\partial t_{xy}}{\partial x} + \lambda v\,\frac{\partial t_{xy}}{\partial y} - a_{xx}\,\frac{\partial v}{\partial x} - a_{yy}\,\frac{\partial u}{\partial y} + t_{xy} = 0.$$

The only difference with (13) is the addition of the second order terms. The first order velocity terms in (25) are of lower order compared to the second-order terms and should not be taken into account if the type of the PDE's has to be found.

Following Courant and Hilbert ([4] pp. 180) we try to find the characteristic equation by replacing (25) by a first-order system of differential equations. Therefore, we introduce

$$q = \frac{\partial u}{\partial x}, \; r = \frac{\partial u}{\partial y}, \; s = \frac{\partial v}{\partial x}, \; w = \frac{\partial v}{\partial y}, \tag{26}$$

with the extra compatibility equations:

$$\frac{\partial v}{\partial x} - \frac{\partial q}{\partial y} = 0, \tag{27}$$

$$\frac{\partial w}{\partial x} - \frac{\partial s}{\partial y} = 0, \tag{28}$$

For equation $(25)_3$ we get:

$$q + w = 0, \tag{29}$$

hence the variable w can be eliminated from the equations. Note, that for the Maxwell models the continuity equation is a partial differential equation contributing to the type and for the Jeffreys models (and the Navier-Stokes equation as well) it only acts as a constraint on lower-order terms.

Introduction of (26) results in the following system of equations:

$$\begin{aligned}
&\frac{\partial p}{\partial x} - \eta_s \frac{\partial q}{\partial x} - \eta_s \frac{\partial v}{\partial y} - \frac{\partial t_{xx}}{\partial x} - \frac{\partial t_{xy}}{\partial y} = \ldots, \\
&\frac{\partial p}{\partial y} - \eta_s \frac{\partial s}{\partial x} + \eta_s \frac{\partial q}{\partial y} - \frac{\partial t_{xy}}{\partial x} - \frac{\partial t_{yy}}{\partial y} = \ldots, \\
&\frac{\partial r}{\partial x} - \frac{\partial q}{\partial y} = 0, \\
&- \frac{\partial q}{\partial x} - \frac{\partial s}{\partial y} = 0, \\
&\lambda \frac{\partial t_{xx}}{\partial t} + \lambda u \frac{\partial t_{xx}}{\partial x} + \lambda v \frac{\partial t_{xx}}{\partial y} = \ldots, \\
&\lambda \frac{\partial t_{yy}}{\partial t} + \lambda u \frac{\partial t_{yy}}{\partial x} + \lambda v \frac{\partial t_{yy}}{\partial y} = \ldots, \\
&\lambda \frac{\partial t_{xy}}{\partial t} + \lambda u \frac{\partial t_{xy}}{\partial x} + \lambda v \frac{\partial t_{xy}}{\partial y} = \ldots,
\end{aligned} \tag{30}$$

where the right-hand sides are of lower differential order. We can easily calculate the corresponding characteristic equation from the procedures given by Courant and Hilbert [4]:

$$Q(\phi_i) = \eta_s (\lambda\xi)^3 (\phi_x^2 + \phi_y^2)^2 = 0, \tag{31}$$

where $\xi = \phi_t + u\phi_x + v\phi_y = \dot{\phi}$.

From (30) we see that the velocity-gradient terms are of lower order. The difference between the several Maxwell models is confined to these terms. Therefore, equation (31) is also valid for Jeffreys-type models where $\underline{\underline{t}}'$ in (8) satisfies other Maxwell-type equations.

The first factor in (31) yields:

$$\lambda\xi = \lambda(\phi_t + u\phi_x + v\phi_y) = 0. \tag{32}$$

This is exactly the same as for the Maxwell-type models, except that the characteristic surfaces of particle trajectories in the x, y, t -space are now of multiplicity three, as could be expected because of the number of stress components. For the Maxwell models particle trajectories are only double characteristics, as we have discussed above. For the steady case again the streamlines are characteristics.

The second factor in (31) leads to the conclusion that for the unsteady flows the x-y plane is a double characteristic surface. This factor represents then a parabolic type. For the steady case, however, no real characteristics exist and the factor represents an elliptic type.

For a generalized Jeffreys-type model (e.g. the Leonov-model given by (9) and (10)) the characteristic equation for plane flow becomes:

$$Q(\phi_i) = -\eta_s(\theta_1\xi)^3 (\theta_2\xi)^3 \ldots (\theta_n\xi)^3 (\phi_x^2 + \phi_y^2)^2 = 0. \tag{33}$$

For every mode there is a factor $(\theta_k\xi)^3$, representing the particle trajectories as characteristics of multiplicity three, corresponding to three stress components for mode k.

BOUNDARY CONDITIONS FOR STEADY FLOW

The boundary conditions for the PDE's describing the steady flow of Maxwell-type and Jeffreys-type models are normally chosen intuitively, e.g. specification of the velocities on the whole boundary, and specification of the extra stresses on the inflow boundary to represent fluid memory. In the report by Hulsen [20] is shown that these boundary conditions are not correct for the UCM model, meaning that the problem is not well-posed. For a linearized UCM model Hulsen shows that only two stresses can be prescribed at the inflow boundary. As the discussion requires much space it is omitted here.

It is assumed, that the correct boundary conditions for steady flow may be found considering only the higher-order terms. In the case of a Jeffreys-type model, the higher-order terms are given by the left-hand side of (30), with $\frac{\partial}{\partial t} = 0$. The characteristic equation is given by (31). It is easily shown, that $(30)_{5,6,7}$ are actually the normal forms belonging to the eigen vectors following from the real characteristic values [20]. In order to integrate these equations, the stresses have to be specified at the boundary, having ingoing characteristics (Γ^+).

If we assume, that the extra stresses have been calculated, the resulting equations in (30) have the same form as the Navier-Stokes equation with a viscosity η_s. This means, that exactly the same boundary conditions for $\underline{v}$ and p have to be specified as in the Navier-Stokes case (e.g. $\underline{v}$ specified on Γ and a level given for p).

For a model having multiple modes, all the extra-stress components have to be specified on Γ^+ in the steady case, to represent the fluid memory.

In this paper, we only consider discontinuities in the flow of Jeffreys-type models. For a discussion of jump conditions for the UCM model we refer to [20].

The steady momentum equation follows from (1) and (8) and can be rewritten in the following form:

$$\mathrm{div}(\rho \underline{v}\underline{v} + p\underline{\underline{1}} - \underline{\underline{t}}) - \eta_s \Delta \underline{v} - \rho \underline{f} = \underline{0}, \tag{34}$$

where the continuity equation (2) has been used. We try to find a weak solution of (34) and multiply this equation by a testfunction $\underline{\Psi}$ and integrate over the domain Ω:

$$\int_\Omega [\mathrm{div}(\rho \underline{v}\underline{v} + p\underline{\underline{1}} - \underline{\underline{t}}) - \eta_s \Delta \underline{v} - \rho \underline{f}] \cdot \underline{\Psi} d\Omega = 0, \tag{35}$$

for every testfunction $\underline{\Psi}$. Using Gauss' theorem yields:

$$-\int_\Omega (\rho \underline{v}\underline{v} + p\underline{\underline{1}} - \underline{\underline{t}})^T : \mathrm{grad}\underline{\Psi} d\Omega - \int_\Omega \eta_s \underline{v} \cdot \Delta \underline{\Psi} d\Omega +$$

$$- \int_\Omega \rho \underline{f} \cdot \underline{\Psi} d\Omega = 0 \; . \tag{36}$$

A solution that satisfies (36) is called a generalized or weak solution.

Let $C : \phi(x, y) = 0$ be a curve of discontinuity, which divides Ω into two domains where $\underline{v}$ and p are smooth and satisfy (34). Applying Gauss' theorem separately to each domain we find that

$$-\int_C \underline{n} \cdot ([\rho \underline{v}\underline{v} + p\underline{\underline{1}} - \underline{\underline{t}}] + \eta_s [\frac{\partial \underline{v}}{\partial n}]) \cdot \underline{\Psi} dC +$$

$$- \eta_s \int_C \frac{\partial \underline{\Psi}}{\partial n} \cdot [\underline{v}] dC = 0, \tag{37}$$

where [ ] denotes a jump across C and $\underline{n}$ is a vector normal to C. This is valid for every testfunction $\underline{\Psi}$ and we get the following jump conditions:

$$[\underline{v}] = 0, \tag{38}$$

$$\underline{n} \cdot [p\underline{\underline{1}} - \underline{\underline{t}}] = \eta_s [\frac{\partial \underline{v}}{\partial n}]. \tag{39}$$

Note that (38) implies that the continuity equation does not give extra jump conditions.

Because of the continuity of the velocity field, it is sufficient to write the constitutive equation of the Jeffreys-type model in the following form:

$$\mathrm{div}(\lambda \underline{n}\underline{\underline{t}}) + \underline{\underline{g}}(\underline{\underline{h}}, \underline{\underline{t}}) = \underline{\underline{0}}, \tag{40}$$

where the function $\underset{\approx}{g}$ is integrable throughout the domain Ω. The jump conditions for (40) are easily derived:

$$v_n[\underline{\underline{t}}] = \underline{\underline{0}}, \tag{41}$$

where $v_n = \underline{n}.\underline{v}$. For $[\underline{\underline{t}}] \neq 0$ we see that $v_n = 0$, which means that the discontinuity surfaces are stream surfaces (=characteristic surfaces).

Now, we introduce the unit base-vectors $(\underline{e}_1, \underline{e}_2, \underline{n})$ such that $\underline{e}_1$ and $\underline{e}_2$ are tangential and $\underline{n}$ normal to the discontinuity surface. The velocity is continuous. This means that there are no jumps in the tangential derivatives:

$$\left[\frac{\partial \underline{v}}{\partial x_1}\right] = \left[\frac{\partial \underline{v}}{\partial x_2}\right] = \underline{0}, \tag{42}$$

where x_1, x_2 are the tangential coordinates along $\underline{e}_1$, $\underline{e}_2$.

From the continuity equation (2) follows:

$$[\mathrm{div}\ \underline{v}] = \left[\underline{n} \cdot \frac{\partial \underline{v}}{\partial n} + \underline{e}_1 \cdot \frac{\partial \underline{v}}{\partial x_1} + \underline{e}_2 \cdot \frac{\partial \underline{v}}{\partial x_2}\right] =$$

$$= \underline{n} \cdot \left[\frac{\partial \underline{v}}{\partial n}\right] = 0, \tag{43}$$

where (42) has been used. Using these equations, we finally get the jump conditions for the Jeffreys model:

$$[\underline{v}] = 0,$$

$$[p] = [t_{nn}],$$

$$[\underline{\tau}] = \eta_s\left[\frac{\partial \underline{v}}{\partial n}\right], \qquad \left(\underline{n} \cdot \left[\frac{\partial \underline{v}}{\partial n}\right] = 0\right),$$

$$v_n[\underline{\underline{t}}] = \underline{\underline{0}}, \tag{44}$$

where $t_{nn} = \underline{n}.\underline{\underline{t}}.\underline{n}$, the normal stress component and $\underline{\tau} = \underline{n}.\underline{\underline{t}} - t_{nn}\underline{n}$, the tangential stress vector. Note that all stress components of $\underline{\underline{t}}$ are allowed to jump, in contrast to the Maxwell model where $[\underline{\tau}]$ must be continuous [20]. This jump is compensated by a jump in the velocity gradient (= shear rate) to achieve local equilibrium. The magnitudes of the jumps are imposed by the boundary conditions.

We expect that for smooth boundaries and smooth initial and/or boundary conditions, the solution will be smooth as well. It is shown in [20] for the convection equation that if sharp corners are present on the boundary, it is possible to have discontinuities depending on the nature of the singularity in the solution. A similar situation arises for viscoelastic flows in domains with complex boundaries and it is possible that discontinuities develop from sharp corners. Moreover, for viscoelastic fluids there is an inherent discontinuity of stresses on the boundary at a sharp corner [19]. This may also effect possible discontinuities in the flow domain.

COMPUTATIONAL SCHEMES AND DISCUSSION OF RESULTS

Various computational schemes have been proposes for the solution of viscoelastic flow problems. Finite difference methods and finite element methods for primitive variables $\underline{v}$, p, $\underline{\underline{t}}$ as well as for streamfunction-vorticity formulations. On the other side particle-tracking methods have been used for models in integral form. For a survey see [5].

In view of the analysis presented in this paper, it is not surprising that methods that do not discriminate between the elliptic and hyperbolic parts of the set of equations, fail for increasing value of the elasticity in the flow. Remarkable is however, that the particle-tracking methods for integral models also failed for increasing value of the elasticity. It is obvious from our analysis that the upper-convected Maxwell model cannot be used because of problems with the boundary conditions: the number of real characteristics (particle trajectories) of the overall system is not equal to the number of stress components. Both the UCM model and the Jeffreys model based on the UCM model have to be rejected because of the singularity of its elongational viscosity, although in the latter case the number of real characteristics (particle trajectories) is equal to the number of stress components and the correct boundary conditions are in agreement with physics.

Few researchers succeed in computing complex flows up to high values of the elasticity , e.g. Upadhyay and Isayev [21] and Luo and Tanner [22,23]. In [21] a Picard iteration scheme is used: solution of the balance of momentum equations with the continuity equation for estimated stress. Then computation of the stresses is done for the given flow field by integrating the constitutive eqautions along the streamlines. Our analysis shows why this can be successful. We were successful in calculating complex flows with a similar scheme as will be reported in a Master's Thesis by Slikkerveer [24].

Luo and Tanner [22] also developed a finite element scheme that solves the continuity equation and the balance of momentum for estimated extra stresses, while as second step the constitutive model is evaluated along the streamlines. The method is similar to the scheme in [21] but cannot handle recirculation zones. In [21] a Newtonian model is used in the recirculation zones assuming slow flow to be valid. Our scheme can also solve the equations in recirculation regions.

Luo and Tanner tackle various problems with various models. Like others they found that most models failed for increasing but low values of the elasticity. Disappointing where the results (not reported) of the Leonov model for the case of plane extrusion : the model predicted no swelling at all. However, they did not give details which formulation of the Leonov model was used. In our experiments there was a big difference between the properties of the model for one and two relaxation times. However, we did not yet compute the die swell problem.

In a second paper Luo and Tanner [23] used integral models with the same solution scheme, i.e. they computed the stresses from the integral formulation along the streamlines. For the standard models the results were similar to those with the differential form: no way to compute the flow for high valus of the elasticity in the flow. They report one exception: a specific version of the KBKZ model with multiple relaxation times. With this model solutions could be obtained for really high values of the elasticity. Although there were some minor drawbacks of the model in describing correctly physical behaviour of real fluids, the results are very promising.

From our analysis and the numerical experiences that have been cited above can be concluded that for viscoelastic flow to be computable at low as well as high values of elasticity, the following conditions have to be satisfied:

i) The numerical method of solution has to discriminate between the elliptic, parabolic, and hyperbolic parts of the total set of equations. The most simple models to deal with are those for which the real characteristics of the total system coincide with those of the constitutive equations for given velocity field. Then the elliptic/parabolic part of the total system coincides with the equations for mass conservation and balance of momentum for given stress field. A Picard iteration can then be used.

ii) The model has to be physically real and stable, e.g. its shear and and elongational viscosity have to be regular and insensitive to small differences in flow conditions. For a correct deformation history and appropriate boundary conditions, the number of real characteristics of the total system of equations that coincide with particle trajectories should be equal to the number of stress components.

REFERENCES

[1] M. Ahrens, D.D. Joseph, M. Renardy and Y. Renardy (1984): "Remarks on the stability of viscometric flow", Rheol. Acta 23, 345-354.

[2] G. Astarita, and G. Marrucci (1974): "Principles of Non-Newtonian Fluid Mechanics", McGraw-Hill, London.

[3] R.B. Bird, R.C. Armstrong and O. Hassager (1977): "Dynamics of Polymer Liquids", Vol. I, John Wiley, New York.

[4] R. Courant and D. Hilbert (1962): "Methods of Mathematical Physics", Vol. 2, Interscience, New York.

[5] M.J. Crochet, A.R. Davies and K. Walters (1984): "Numerical Simulation of Non-Newtonian flow", Elsevier, Amsterdam.

[6] F. Dupret, J.M. Marchal and M.J. Crochet (1985): "On the consequence of discretization errors in the numerical calculation of visco-elastic flow", J. Non-Newt. Fluid Mech. 18, 173-186.

[7] H. Janeschitz-Kriegl (1983): "Polymer Melt Rheology and Flow Birefringence", Springer-Verlag, Berlin.

[8] D.D. Joseph, M. Renardy and J.C. Saut (1985): "Hyperbolicity and change of type in the flow of viscoelastic fluids", Arch. Rat. Mech. Anal. 87, 213-251.

[9] A.I. Leonov (1976): "Nonequilibrium thermodynamics and rheology of viscoelastic polymer media", Rheol. Acta 15, 85-98.

[10] M.Luskin (1984): "On the classification of some equations for viscoelasticity", J. Non-Newt. Fluid Mech. 16, 3-11.

[11] H.K. Moffatt (1964): "Viscous and resistive eddies near a sharp corner", J. Fluid Mech. 18, 1-18.

[12] S.A. Regirer and I.M. Rutkevich (1968): "Certain singularities of the hydrodynamic equations of non-Newtonian media", PMM 32, 942-945.

[13] I.M. Rutkevich (1969): "Some general properties of the equations of viscoelastic incompressible fluid dynamics", PMM 33, 42-51.

[14] I.M. Rutkevich (1970): "The propagation of small perturbations in a viscoelastic fluid", PMM 34, 41-56.
[15] C. Truesdell (1977): "A First Course in Rational Continuum Mechanics", Vol. I, Academic Press, New York.
[16] J.S. Ultman and M.M. Denn (1970): "Anamolous heat transfer and a wave phenomenon in dilute polymer solutions", Trans. Soc. Rheol. 14, 307-317.
[17] J.Y. Yoo, M. Ahrens and D.D. Joseph (1985): "Hyperbolicity and change of type in sink flow", J. Fluid Mech. 153, 203-214.
[18] J.Y. Yoo and D.D. Joseph (1985): "Hyperbolicity and change of type in the flow of viscoelastic fluids through channels", J.Non-Newt. Fluid Mech. 19, 15-41.
[19] J. van der Zanden, G.D.C. Kuiken, A. Segal, W.J. Lindhout and M.A. Hulsen (1984): "Numerical experiments and theoretical analysis of the flow of an elastic liquid of the Maxwell-Oldroyd type in the presence of geometrical singularities", Dept. of Mech. Eng., Delft University of Technology, WTHD report no. 164.
Similar to: Applied Scientific Research 42 (1985) 303-318.
[20] M.A. Hulsen (1986) : "Analysis of the equations for viscoelastic flow: type, boundary conditions and discontinuities.", Dept. of Mech. Eng., Delft University of Technology, WTHD report.
[21] R.K. Upadhyay and A.I. Isayev (1986): "Simulation of two-dimensional planar flow of viscoelastic fluid", Rheol. Acta 25, 80-94.
[22] X.L. Luo and R.I. Tanner (1986): "A streamline scheme for solving viscoelastic flow problems. Part I: Differential constitutive equations", J.Non-Newt. Fluid Mech. 21, 179-199.
[23] X.L. Luo and R.I. Tanner (1986): "A streamline scheme for solving viscoelastic flow problems. Part II: Integral constitutive models", J.Non-Newt. Fluid Mech. 22, 61-89.
[24] P.J. Slikkerveer, Master's Thesis, To be published in 1987.

Multigrid and Defect Correction for the Efficient Solution of the Steady Euler Equations

Barry Koren, Stefan Spekreijse
Centre for Mathematics and Computer Science
P.O. Box 4079, 1009 AB Amsterdam, The Netherlands

An efficient iterative solution method for second-order accurate discretizations of the 2D Steady Euler equations is described and results are shown. The method is based on a nonlinear multigrid method and on the defect correction principle. Both first- and second-order accurate finite-volume upwind discretizations are considered. In the second-order discretization a limiter is used.

An Iterative Defect Correction process is used to approximately solve the system of second-order discretized equations. In each iteration of this process, a solution is computed of the first-order system with an appropriate right-hand side. This solution is computed by a nonlinear multigrid method, where Symmetric Gauss-Seidel relaxation is used as the smoothing procedure.

The computational method does not require any tuning of parameters. Flow solutions are presented for an airfoil and a bi-airfoil with propeller disk. The solutions show good resolution of all flow phenomena and are obtained at low computational cost. Particularly with respect to efficiency, the method contributes to the state of the art in computing steady Euler flows with discontinuities.

1980 Mathematics Subject Classification: 35L65, 35L67, 65N30, 76G15, 76H05.
Key Words and Phrases: steady Euler equations, multigrid methods, defect correction.
Note: This work was supported in part by the Netherlands Technology Foundation (STW).

1. Introduction

The Euler equations describe compressible inviscid gas flows with rotation. They are widely used in the aerospace industry. The Euler equations are derived by considering the laws of conservation of mass, momentum and energy for an inviscid gas. The result is a nonlinear hyperbolic system of conservation laws. Only for very simple flow problems, analytical solutions exist. For almost all engineering problems solutions must be found numerically. Several discretization methods have been developed which yield solutions of good quality (good resolution of shock waves, slip lines, etc.). However, generally the computational cost is high. In 1983 a project was started at the Centre for Mathematics and Computer Science (CWI) in Amsterdam for the development of more efficient methods. So far, a multigrid method for the solution of the 2D steady Euler equations has been developed, implemented and tested.

In the method, the steady Euler equations are discretized by a finite-volume upwind discretization [9]. Both first- and second-order discretizations are obtained by the projection-evolution approach [14]. In the projection-stage of this approach the discrete values, located in the volume centers, are interpolated to yield continuous distributions in each volume. First-order accuracy is obtained by piecewise constant interpolation, second-order accuracy by piecewise linear interpolation. In case of flows with discontinuities (shock waves or slip lines), the occurrence of spurious non-monotonicity (wiggles) when using a second-order interpolation, is suppressed by the use of a limiter in the interpolation formulae [23]. In this paper we use the Van Albada limiter [1,20]. In the evolution-stage, a Riemann problem is considered for the computation of the flux at each volume wall. To approximately solve each Riemann problem we use the Osher scheme [16].

To obtain solutions of the system of first-order discretized equations, the nested nonlinear multigrid (FMG-FAS-) iteration method appears to be a very efficient solution method [9]. However, the multigrid solution of a system of second-order discretized equations appears to be less efficient [21].

Therefore, for the solution of the latter system we make use of an Iterative Defect Correction (IDeC-) process [2]. In each iteration of this process, the second-order discretization is only used for the construction of an appropriate right-hand side for a system of first-order discretized equations. Nonlinear multigrid (FAS) is used to solve this system.

To show that the method efficiently yields good Euler flow solutions we consider a NACA0012-airfoil at:

(i) $M_\infty=0.85$, $\alpha=1^0$ (flow with upper surface shock, lower surface shock and tail slip line), and
(ii) $M_\infty=1.2$, $\alpha=7^0$ (flow with detached bow shock, oblique tail shock and tail slip line).

Both flows are isenthalpic. To show the use of the method in case of a non-isenthalpic flow, we consider a NACA0012-bi-airfoil with working propeller disk, at $M_\infty=0.5$, $\alpha=2^0$ (flow with internal shock and subsonic jet).

In section 2 a description is given of the first- and second-order discretizations. In section 3 the solution method is described. In section 4 we discuss the numerical results, and in section 5 some conclusions are listed.

2. Discretization

Consider on an open domain $\Omega\in\mathbb{R}^2$ the 2D steady Euler equations in conservation form and without source terms:

$$\frac{\partial f(q)}{\partial x}+\frac{\partial g(q)}{\partial y} = 0\,, \tag{2.1}$$

where $q=(\rho,\rho u,\rho v,E)^T$ is the state vector of conservative variables, and where $f(q)=(\rho,\rho u^2+p,\rho uv,(E+p)u)^T$ and $g(q)=(\rho v,\rho uv,\rho v^2+p,(E+p)v)^T$ are the flux vectors. The so-called primitive variables of (2.1) are the density ρ, the velocity components u and v, and the pressure p. For a perfect gas, the total energy per unit of volume, E, is related to the primitive variables as $E = p/(\gamma-1)+\frac{1}{2}\rho(u^2+v^2)$ where γ is the ratio of specific heats.

To allow solutions with discontinuities we consider the Euler equations in their integral form. Then the 2D steady Euler equations read

$$\int_{\partial\Omega^*}\{\cos\phi f(q)+\sin\phi g(q)\}ds = 0\,,\quad \forall\Omega^*\subset\Omega, \tag{2.2}$$

where $\Omega^*\subset\Omega$ is an arbitrary subregion of Ω, $\partial\Omega^*$ the boundary of Ω^*, and $(\cos\phi,\sin\phi)$ the outward unit normal on $\partial\Omega^*$. A straightforward and simple discretization of (2.2) is obtained by subdividing Ω into disjunct quadrilateral subregions $\Omega_{i,j}$ (the finite volumes) and by requiring that

$$\int_{\partial\Omega_{i,j}}(\cos\phi f(q)+\sin\phi g(q))ds = 0 \tag{2.3}$$

for each volume $\Omega_{i,j}$ separately. We restrict ourselves to subdivisions such that only $\Omega_{i,j\pm1}$ and $\Omega_{i\pm1,j}$ are the neighbouring volumes of $\Omega_{i,j}$.

Using the rotational invariance of the Euler equations:

$$\cos\phi f(q)+\sin\phi g(q) = T^{-1}(\phi)f(T(\phi)q), \tag{2.4}$$

where $T(\phi)$ is the rotation matrix

$$T(\phi) = \begin{bmatrix} 1 & 0 & 0 & 0\\ 0 & \cos\phi & \sin\phi & 0\\ 0 & -\sin\phi & \cos\phi & 0\\ 0 & 0 & 0 & 1\end{bmatrix}, \tag{2.5}$$

(2.3) becomes:

$$\int_{\partial\Omega_{i,j}} T^{-1}(\phi) f(T(\phi)q) ds = 0 . \tag{2.6}$$

A numerical approximation of this formula is obtained by

$$F_{i,j} := f_{i+\frac{1}{2},j} + f_{i,j+\frac{1}{2}} - f_{i-\frac{1}{2},j} - f_{i,j-\frac{1}{2}} = 0, \tag{2.7}$$

with

$$f_{i+\frac{1}{2},j} = l_{i+\frac{1}{2},j} T^{-1}(\phi_{i+\frac{1}{2},j}) f_R(T(\phi_{i+\frac{1}{2},j}) q^l_{i+\frac{1}{2},j}, T(\phi_{i+\frac{1}{2},j}) q^r_{i+\frac{1}{2},j}) \tag{2.8}$$

and similar relations for $f_{i-\frac{1}{2},j}$ and $f_{i,j\pm\frac{1}{2}}$. In (2.8), $l_{i+\frac{1}{2},j}$ is the length of the volume wall $\partial\Omega_{i+\frac{1}{2},j} = \Omega_{i,j} \cap \Omega_{i+1,j}$ and $(\cos\phi_{i+\frac{1}{2},j}\ ,\ \sin\phi_{i+\frac{1}{2},j})$ the outward unit normal on $\partial\Omega_{i+\frac{1}{2},j}$ (fig. 2.1a). Further, $f_R : \mathbb{R}^4 \times \mathbb{R}^4 \rightarrow \mathbb{R}^4$ is a so-called approximate Riemann-solver and $q^l_{i+\frac{1}{2},j}$ and $q^r_{i+\frac{1}{2},j}$ are state vectors located at the left and right side of volume wall $\partial\Omega_{i+\frac{1}{2},j}$ (fig. 2.1b). The flux vector $f_{i+\frac{1}{2},j}$ represents the transport of mass, momentum and energy per unit of time, across $\partial\Omega_{i+\frac{1}{2},j}$. For a more detailed discussion of (2.7) and (2.8) we refer to [9,19].

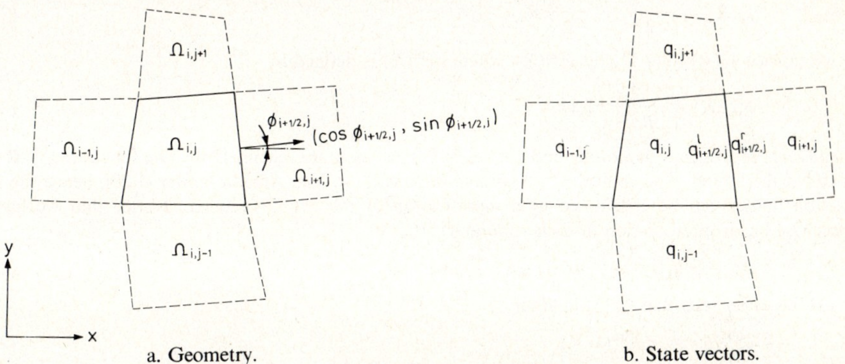

a. Geometry. b. State vectors.

Fig. 2.1: Finite volume $\Omega_{i,j}$.

The application of an approximate Riemann-solver is the essential part of the evolution-stage, whereas the computation of the states $q^l_{i+\frac{1}{2},j}$ and $q^r_{i+\frac{1}{2},j}$ is the essential part of the projection-stage [14]. As the name suggests an approximate Riemann-solver is used to obtain an approximate solution of the Riemann initial-value problem [4,6]. Several approximate Riemann-solvers exist [13,16,18,22]. Here, we use Osher's Riemann-solver because of its consistent treatment of boundary conditions and its continuous differentiability [9,16,17]. For details about an efficient implementation of Osher's approximate Riemann-solver we refer to [9].

Depending on the way the states $q^l_{i+\frac{1}{2},j}$ and $q^r_{i+\frac{1}{2},j}$ are computed, the discretization (2.7) is first- or second-order accurate. First-order accuracy is obtained by taking

$$\begin{aligned} q^l_{i+\frac{1}{2},j} &= q_{i,j}, \text{ and} \\ q^r_{i+\frac{1}{2},j} &= q_{i+1,j}. \end{aligned} \tag{2.9}$$

Second-order accuracy can be obtained by for example the κ-schemes introduced by Van Leer [14]:

$$\begin{aligned} q^l_{i+\frac{1}{2},j} &= q_{i,j} + \frac{1+\kappa}{4}(q_{i+1,j} - q_{i,j}) + \frac{1-\kappa}{4}(q_{i,j} - q_{i-1,j}), \text{ and} \\ q^r_{i+\frac{1}{2},j} &= q_{i+1,j} + \frac{1+\kappa}{4}(q_{i,j} - q_{i+1,j}) + \frac{1-\kappa}{4}(q_{i+1,j} - q_{i+2,j}) , \end{aligned} \tag{2.10}$$

with $\kappa \in [-1,1]$. For $\kappa = -1$, $\kappa = 0$, $\kappa = 1/3$ and $\kappa = 1$ we find respectively: the fully one-sided upwind scheme, the Fromm scheme, the upwind biased scheme (third-order accurate for 1D problems) and the central scheme. A disadvantage of these κ-schemes is that near discontinuities, spurious non-monotonicity (wiggles) appears [11]. A way to avoid this is by using a limiter. We modify the κ-schemes by introducing a limiter such that the schemes become monotone and remain second-order accurate. Let $q^{l\,(k)}_{i+\frac{1}{2},j}$ and $q^{r\,(k)}_{i+\frac{1}{2},j}$ be the kth component ($k = 1,2,3,4$) of $q^l_{i+\frac{1}{2},j}$ and $q^r_{i+\frac{1}{2},j}$. We rewrite (2.10) as

$$\begin{aligned} q^{l\,(k)}_{i+\frac{1}{2},j} &= q^{(k)}_{i,j} + \tfrac{1}{2}\psi_\kappa(R^{(k)}_{i,j})(q^{(k)}_{i,j} - q^{(k)}_{i-1,j}), \text{ and} \\ q^{r\,(k)}_{i+\frac{1}{2},j} &= q^{(k)}_{i+1,j} + \tfrac{1}{2}\psi_\kappa(1/R^{(k)}_{i+1,j})(q^{(k)}_{i+1,j} - q^{(k)}_{i+2,j}), \end{aligned} \tag{2.11}$$

where

$$R^{(k)}_{i,j} = \frac{q^{(k)}_{i+1,j} - q^{(k)}_{i,j}}{q^{(k)}_{i,j} - q^{(k)}_{i-1,j}}, \tag{2.12}$$

and where $\psi_\kappa : \mathbb{R} \to \mathbb{R}$ is defined by

$$\psi_\kappa(R) = \frac{1-\kappa}{2} + \frac{1+\kappa}{2} R. \tag{2.13}$$

If we replace $\psi_\kappa(R)$ in (2.11) by $\psi^{\lim}_\kappa(R)$, where $\psi^{\lim}_\kappa(R)$ is defined by

$$\psi^{\lim}_\kappa(R) = \frac{2R}{R^2+1}\psi_\kappa(R), \tag{2.14}$$

then (2.11) results in a monotone and yet second-order accurate scheme [20]. The function $\psi^{\lim}_\kappa : \mathbb{R} \to \mathbb{R}$ is called the limiter. The choice $\kappa = 0$ corresponds with the Van Albada limiter [1,20], hence the Van Albada limiter can be considered as a modification of the Fromm scheme. Notice that we have a piecewise linear interpolation in each volume if

$$q^l_{i+\frac{1}{2},j} - q_{i,j} = q_{i,j} - q^r_{i-\frac{1}{2},j}. \tag{2.15}$$

It can be easily seen that (2.15) holds if

$$\psi^{\lim}_\kappa(R) = R\psi^{\lim}_\kappa(1/R), \tag{2.16}$$

and this is only true for $\kappa = 0$. An advantage of the Van Albada limiter is that in the neighbourhood of discontinuities the scheme resembles the fully one-sided upwind scheme, which is a natural scheme in such regions. For all flow solutions presented in this paper we used $\psi^{\lim}_0(R)$ although $\psi^{\lim}_{1/3}(R)$ seems a reasonable choice as well.

In case $\Omega_{i,j}$ is a boundary volume, so that for example $\partial\Omega_{i+\frac{1}{2},j}$ is part of the domain boundary, no limiter can be used to compute $q^l_{i+\frac{1}{2},j}$ and $q^r_{i-\frac{1}{2},j}$. In this case we use a simple linear interpolation, i.e.

$$\begin{aligned} q^l_{i+\frac{1}{2},j} &= q_{i,j} + \tfrac{1}{2}(q_{i,j} - q_{i-1,j}), \text{and} \\ q^r_{i-\frac{1}{2},j} &= q_{i,j} - \tfrac{1}{2}(q_{i,j} - q_{i-1,j}). \end{aligned} \tag{2.17}$$

The boundary conditions, together with the state $q^l_{i+1/2,j}$, are used to compute the state $q^r_{i+\frac{1}{2},j}$. This computation is done by considering the Riemann initial-boundary value problem [9,17]. The flux $f_{i+\frac{1}{2},j}$ at $\partial\Omega_{i+\frac{1}{2},j}$ is computed by (2.8).

3. Solution method

The method to solve the system of nonlinear discretized equations is based on a multigrid technique. For readers unfamiliar with multigrid techniques we refer to [3,5].

Let

$$F^1_h(q_h) = r_h, \tag{3.1}$$

and

$$F_h^2(q_h) = r_h \tag{3.2}$$

be first- and second-order accurate finite-volume upwind discretizations of the 2D steady Euler equations with source term r. Hence, $(F_h^1(q_h))_{i,j}=F_{i,j}$ is defined by (2.7), (2.8) and (2.9), and $(F_h^2(q_h))_{i,j}=F_{i,j}$ is defined by (2.7), (2.8), (2.11) and (2.12) with $\psi_\kappa(R)=\psi_0^{\lim}(R)$ (the Van Albada limiter). Although in general $r = 0$, we prefer to describe the solution method for systems with an arbitrary right-hand side. The subscript h denotes the meshwidth. To apply multigrid we construct a nested set of grids, such that each volume in a grid is the union of 4 volumes in the next finer grid, in the obvious way. Let Ω_{h_i} with $h_1>h_2>\cdots>h_l = h$ be a sequence of such nested grids. So Ω_{h_1} and Ω_{h_l} are respectively the coarsest and the finest grid.

The solution method for (3.2) can be divided into three successive stages. The first stage is the Full Multigrid (FMG-) method, which is used to find a good initial approximation of (3.1). The second stage is a nonlinear multigrid (FAS-) iteration method, which is used to find a better approximate solution of (3.1). The first iterand is the solution obtained by the FMG-method. The third and last stage is an Iterative Defect Correction (IDeC-) process, which is used to find an approximate solution of (3.2). The first iterand of this process is obtained from the second stage. We will now discuss these stages more fully.

Stage I : The Full Multigrid (FMG-) method.
Let

$$F_{h_i}^1(q_{h_i}) = r_{h_i} \tag{3.3}$$

be the first-order discretization on Ω_{h_i}, $i = 1,2,...,l$. The FMG-method (or nested iteration) starts with a crude initial exstimate of q_{h_1}; the solution on the coarsest grid. To obtain an initial estimate on the finer grid $\Omega_{h_{i+1}}$, first the solution on the next coarser grid Ω_{h_i} is improved by a single FAS-iteration (stage II). Hereafter this improved approximation is interpolated to the finer grid $\Omega_{h_{i+1}}$. These steps are repeated until the highest level has been reached. The interpolation used to obtain the first guess on a next finer grid is a bilinear interpolation. For this purpose the grid Ω_{h_i} is subdivided into disjunct sets of 2×2 volumes. The four states corresponding with each set are interpolated in a bilinear way, and since each volume of Ω_{h_i} overlaps 2×2 finer grid volumes of $\Omega_{h_{i+1}}$, 4×4 new states are obtained on $\Omega_{h_{i+1}}$.

Stage II: The nonlinear multigrid (FAS-) iteration method.
To find a better approximation to (3.1) we apply the FAS-iteration method on the finest grid (Ω_{h_l}). One FAS-iteration on a general grid Ω_{h_i} is recursively defined by the following steps:

(0) Start with an approximate solution of q_{h_i}.
(1) Improve q_{h_i} by application of p (pre-) relaxation iterations to $F_{h_i}^1(q_{h_i}) = r_{h_i}$.
(2) Compute the defect $d_{h_i} := r_{h_i} - F_h^1(q_{h_i})$.
(3) Find an approximation of $q_{h_{i-1}}$ on the next coarser grid $\Omega_{h_{i-1}}$. Either use $q_{h_{i-1}} := \hat{I}_{h_i}^{h_{i-1}} q_{h_i}$, where $\hat{I}_{h_i}^{h_{i-1}}$ is a restriction operator, or use the last obtained approximation $q_{h_{i-1}}$. (We use this last obtained approximation.)
(4) Compute $r_{h_{i-1}} := F_{h_{i-1}}^1(q_{h_{i-1}}) + I_{h_i}^{h_{i-1}} d_{h_i}$ where $I_{h_i}^{h_{i-1}}$ is another restriction operator.
(5) Approximate the solution of $F_{h_{i-1}}^1(q_{h_{i-1}}) = r_{h_{i-1}}$ by σ FAS- iterations on $\Omega_{h_{i-1}}$. The result is called $\tilde{q}_{h_{i-1}}$. ($\sigma = 1$ results in a so-called V-cycle and $\sigma = 2$ in a W-cycle.)
(6) Correct the current solution by $q_{h_i} := q_{h_i} + I_{h_{i-1}}^{h_i}(\tilde{q}_{h_{i-1}} - q_{h_{i-1}})$, where $I_{h_{i-1}}^{h_i}$ is a prolongation operator.
(7) Improve q_{h_i} by application of q (post-) relaxation iterations to $F_{h_i}^1(q_{h_i}) = r_{h_i}$.

The steps (2) - (6) are called the coarse-grid correction. These steps are skipped on the coarsest grid.

In order to complete the description of a FAS-iteration we have to discuss: (i) the choice of the transfer operators $I^{h_l}_{h_{l-1}}$, $I^{h_{l-1}}_{h_l}$ and $\hat{I}^{h_{l-1}}_{h_l}$, (ii) the relaxation method, and (iii) the FAS-strategy, i.e. the numbers p, q and σ.

(i) Choice of the operators:

The restriction operators $\hat{I}^{h_{l-1}}_{h_l}$ and $I^{h_{l-1}}_{h_l}$ are defined by

$$(q_{h_{l-1}})_{i,j} = (\hat{I}^{h_{l-1}}_{h_l} q_{h_l})_{i,j} := \tfrac{1}{4}\{(q_{h_l})_{2i,2j}+(q_{h_l})_{2i-1,2j}+(q_{h_l})_{2i,2j-1}+(q_{h_l})_{2i-1,2j-1}\}, \text{and} \tag{3.4}$$

$$(r_{h_{l-1}})_{i,j} = (I^{h_{l-1}}_{h_l} r_{h_l})_{i,j} := (r_{h_l})_{2i,2j}+(r_{h_l})_{2i-1,2j}+(r_{h_l})_{2i,2j-1}+(r_{h_l})_{2i-1,2j-1}. \tag{3.5}$$

The prolongation operator $I^{h_l}_{h_{l-1}}$ is defined by

$$(I^{h_l}_{h_{l-1}} q_{h_{l-1}})_{2i,2j}=(I^{h_l}_{h_{l-1}} q_{h_{l-1}})_{2i-1,2j}=(I^{h_l}_{h_{l-1}} q_{h_{l-1}})_{2i,2j-1}=(I^{h_l}_{h_{l-1}} q_{h_{l-1}})_{2i-1,2j-1}:=(q_{h_{l-1}})_{i,j}. \tag{3.6}$$

Note that this prolongation is different from the bilinear interpolation used in FMG. By defining the transfer operators in this way, it can be verified that

$$F^1_{h_{l-1}} = I^{h_{l-1}}_{h_l} F^1_{h_l} I^{h_l}_{h_{l-1}}, \tag{3.7}$$

i.e. the first-order coarse grid discretizations of the steady Euler equations are Galerkin approximations of the fine grid discretizations. This is a very important property because it implies that the coarse grid correction efficiently reduces the smooth component in the residual.

(ii) The relaxation method:

We use Collective Symmetric Gauss-Seidel (CSGS-) relaxation. Collective means that the four variables corresponding to a single volume are relaxed simultaneously. At each volume visited we solve the four nonlinear equations by Newton's method (local linearization). It appears that a single Newton iteration is sufficient. For details about the local linearization formulae we refer to [9].

(iii) The FAS-strategy:

We use a fixed strategy: $\sigma = 1$ and $p = q = 1$, i.e. we use V-cycles with one pre- and one post-relaxation.

Stage III : The Iterative Defect Correction (IDeC-) process.

For an introduction to the defect correction approach we refer to [2]. We approximate the solution of (3.2) with the IDeC-process:

$$F^1_h(q^{(n+1)}_h) = F^1_h(q^{(n)}_h) + (r_h - F^2_h(q^{(n)}_h)) \ , \ n = 0,1,2,... \ , \tag{3.8}$$

where $q^{(0)}_h$ is the solution obtained in stage II. It is clear that the fixed point of this iteration process is the solution of (3.2). In fact it is not really necessary to iterate until convergence. For smooth solutions a single IDeC-iteration is sufficient to obtain second-order accuracy [7]. For solutions with discontinuities experience shows that one or a few IDeC-iterations significantly improve the accuracy of the solution [11].

For each IDeC-iteration we have to solve a first-order system with an appropriate right-hand side. It appeared that it is inefficient to solve this system very accurately. Application of a single FAS-iteration to approximate $q^{(n+1)}_h$ in (3.8) usually is the most efficient strategy [7,11].

In fig. 3.1. we give an illustration of the complete solution process. Suppose there are 5 nested grids ($l = 5$). Between two succeeding points A,B we have one FAS-iteration (V-cycle). Between two succeeding points B,A we have a bilinear interpolation in the FMG-stage, and an appropriate right-hand side computation in the IDeC-stage.

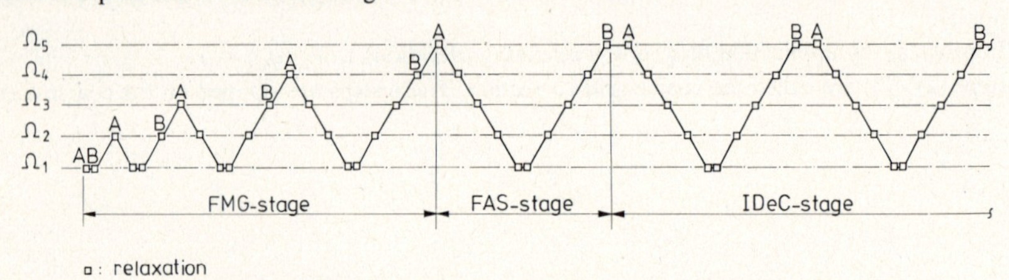

Fig. 3.1: Schematic representation of (a 5-level) solution process.

4. RESULTS

To show that the method is feasible for a good and efficient computation of typical Euler flows, we consider two standard Euler test cases for a NACA0012-airfoil ($M_\infty=0.85$, $\alpha=1°$ and $M_\infty=1.2$, $\alpha=7°$) and compare our results with results from [24]. (M_∞ denotes the Mach number at infinity and α the airfoil's angle of attack.) Both airfoil flows are isenthalpic, i.e. $(E+p)/\rho$ is uniformly constant. (This fact is not exploited in our computations.) To test our method in case of non-isenthalpic Euler flow we consider a configuration composed of two NACA0012-airfoils and a propeller disk. The NACA0012-airfoils are placed in parallel formation, the propeller disk is placed between the airfoil noses (fig. 4.1). The configuration can be interpreted as a model for a bi-plane with airscrew(s) between its leading edges. The propeller disk is modelled as a line-distribution of x-momentum and energy sources. For the NACA0012-bi-airfoil with propeller disk, no results to compare with are available.

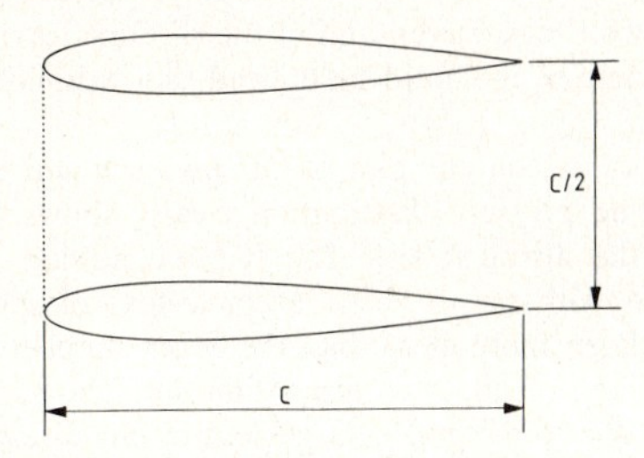

Fig. 4.1: NACA0012-bi-airfoil with propeller disk.

The NACA0012-airfoil:

For the NACA0012-airfoil we use a 128×32 O-type grid with the outer boundary at an approximate distance from the airfoil of 100 chord lengths (fig. 4.2). Following [8,11], we impose unperturbed flow conditions at the outer boundary, although we do not overimpose. For the subsonic outer boundary of the first test case we impose 3 conditions at the inflow part of that boundary ($u=M_\infty \cos\alpha$, $v=M_\infty \sin\alpha$, $c=1$), and 1 condition at the outflow part ($u=M_\infty \cos\alpha$). For the supersonic test case we impose 4 conditions at inflow ($u=M_\infty \cos\alpha$, $v=M_\infty \sin\alpha$, $c=1, p=1$), and nothing at outflow. For both test cases we perform 10 IDeC-iterations and use a multigrid algorithm with 4 coarser grids.

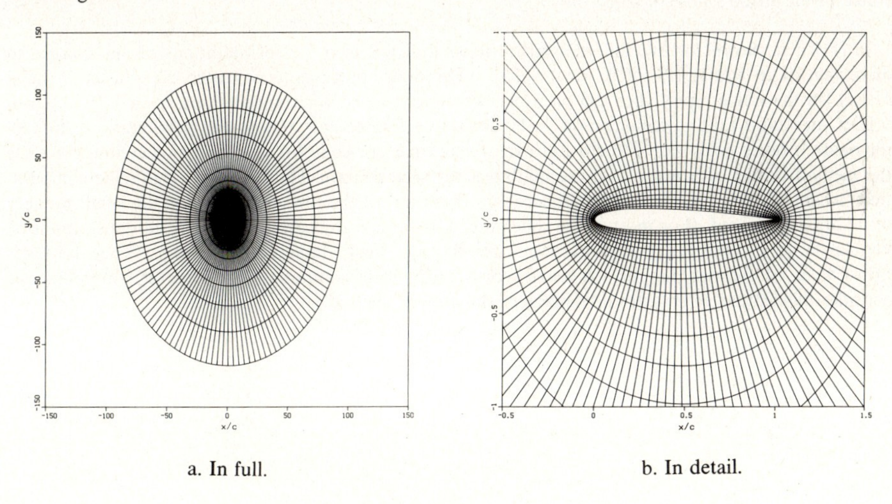

a. In full. b. In detail.

Fig. 4.2: 128×32-grid NACA0012-airfoil.

The results obtained for $M_\infty=0.85$, $\alpha=1°$ are presented in fig. 4.3. In fig. 4.3a and 4.3b we present convergence histories. In fig. 4.3a the residual ratio $\sum_{(i,j)}|(F_h^2(q_h^{(n)}))_{i,j}|/\sum_{(i,j)}|(F_h^2(q_h^{(0)}))_{i,j}|$ (L_1-norm) is plotted versus n; the number of IDeC-iterations. In fig. 4.3b we show the convergence history of the lift and drag force acting on the airfoil. (For a definition of lift, drag and their proper scaling we refer to e.g. [15].) Although the L_1-norm of the residual ratio is decreasing rather slowly, fig. 4.3b shows that a practical convergence of the lift and drag has been obtained after ~7 IDeC-iterations. This is typical for IDeC-processes. The shaded areas in fig. 4.3b represent the values of lift and drag as presented in [24] by 7 other investigators. As the best reference results from [24] we selected those obtained by Schmidt & Jameson. For the lift and drag they find: $c_l=0.3472$, $c_d=0.0557$, whereas we find (after the 10th IDeC-iteration): $c_l=0.3565$, $c_d=0.0582$.

In fig. 4.3c we show a contour plot of the Mach number distribution and make a comparison with the distribution as obtained by Schmidt & Jameson. Both distributions show a good (i.e. a sharp and monotone) capturing of the two shock waves, and of the slip line leaving the airfoil's tail. Concerning the sharpness of the discontinuities, it should be noticed that Schmidt & Jameson used a 320×64 (!) O-type grid.

In fig. 4.3d and 4.3e we show a contour plot of our pressure and entropy distribution. (No reference results are available.) The pressure distribution clearly shows the smoothness of the pressure across the slip line (up to the airfoil's tail). The Kutta-condition is satisfied automatically. The entropy distribution $s/s_\infty-1$, with $s=p\rho^{-\gamma}$ has a convection of spurious entropy generated at the airfoil's nose of 0.003 only. Even more clear than the Mach number distribution, the entropy distribution shows the good capturing of all three discontinuities. The slight spreading of the slip line in downstream direction is only due to the grid enlargement in this direction.

The results obtained for $M_\infty=1.2$, $\alpha=7°$ are presented in fig. 4.4. The convergence histories of the residual ratio and the lift and drag are presented in fig. 4.4a and 4.4b respectively. As the best reference results from [24] we selected for this test case those obtained by Veuillot & Vuillot. As values for the lift and drag they find: $c_l=0.5280$, $c_d=0.1530$, whereas we find (after again the 10th IDeC-iteration): $c_l=0.5237$, $c_d=0.1551$. Veuillot & Vuillot obtained their results on a 201×55 C-type grid.

In fig. 4.4c-4.4e we show the solution obtained for this test case. Clearly visible in the various figures is the detached bow shock, the oblique tail shock, the slip line and the smoothness of the pressure across the latter. The rather large spreading of the bow shock is only due to the rather large grid coarseness in this region (fig. 4.2b). In the entropy distribution (fig. 4.4e) the 'separation' from the bow shock of the contour line $s/s_\infty-1=0.002$ properly shows the variable strength of the bow shock. Just as for the previous test case the entropy distribution has a convection of spurious entropy generated at the airfoil's nose of 0.003 only.

In [11] it is shown for five different airfoil flows that we need 5 IDeC-iterations on an average to drive the lift to within ½% of its final value. (The drag appeared to converge even faster in most cases.) On the single pipe Cyber 205 on which we performed all our computations, for a 128×32-grid, 5 IDeC-iterations take in scalar mode ~100 sec (i.e. ~5 msec per volume and per iteration). In vector mode it takes ~50 sec. We did not extensively tune our code for use on vector computers since the method brings with it some severe inhibiters for vectorization. However, for large scale computations where all data cannot be kept in core, an advantage of the present method is the small number of iterations required. (For most Euler codes this number is significantly larger.) If all data cannot be kept in core, a small number of iterations results in a small data transport load. Since IO-times rather than CPU-times may be the bottleneck in large scale computations on vector computers, we consider this feature as an extra advantage of the present method.

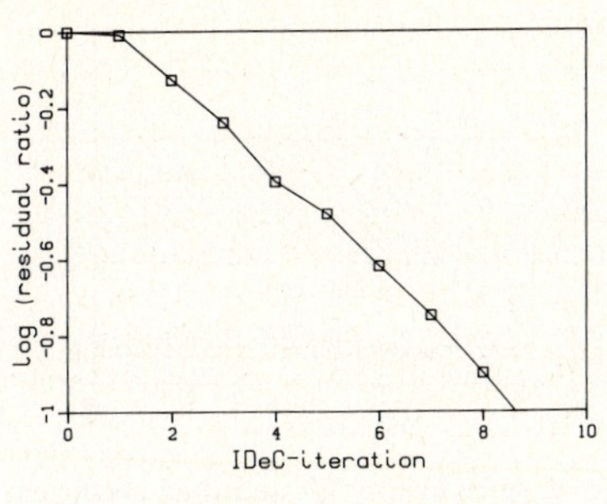

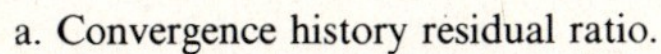

a. Convergence history residual ratio.

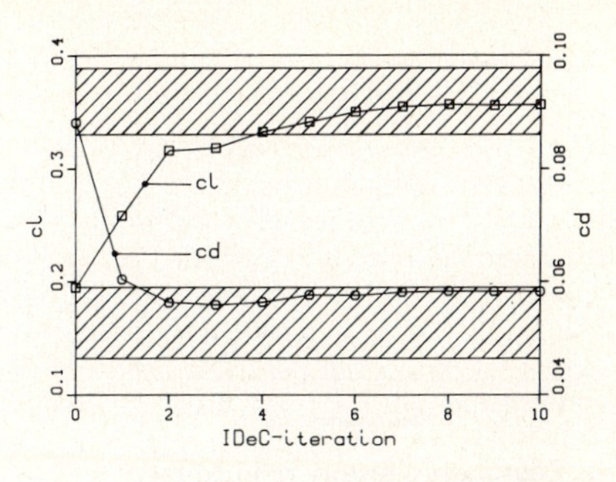

b. Convergence history lift and drag coefficient.

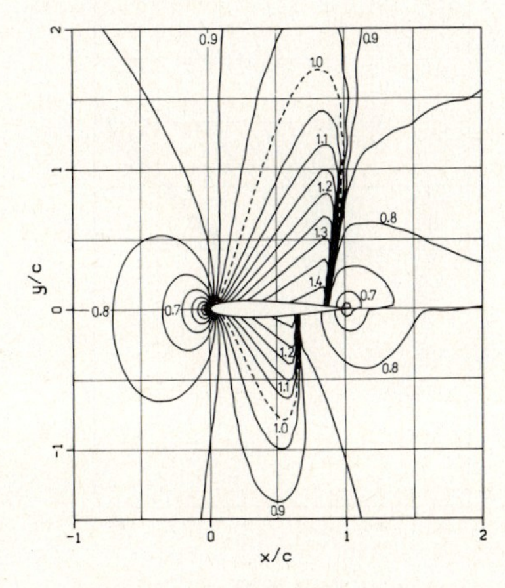

c. Mach number distributions; present result (left) and result Schmidt & Jameson (right).

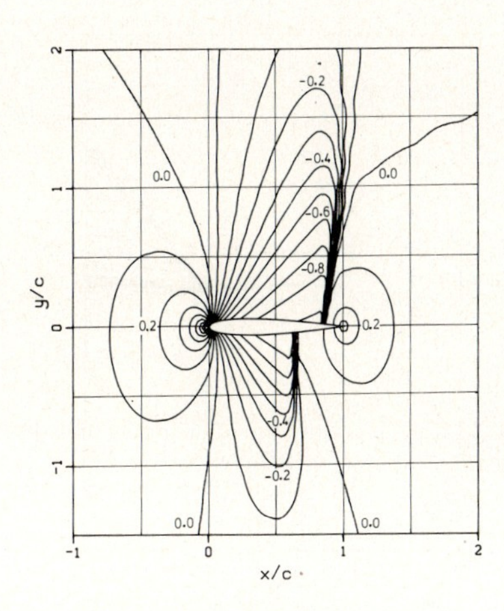

d. Present pressure distribution (c_p).

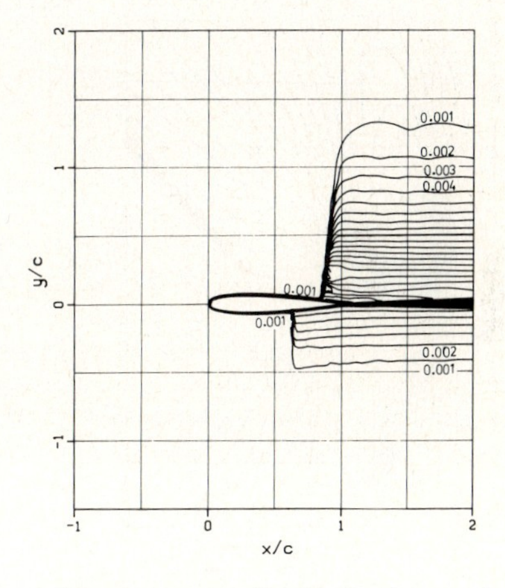

e. Present entropy distribution ($s/s_\infty - 1$).

Fig. 4.3: Results for NACA0012-airfoil at $M_\infty = 0.85$, $\alpha = 1°$.

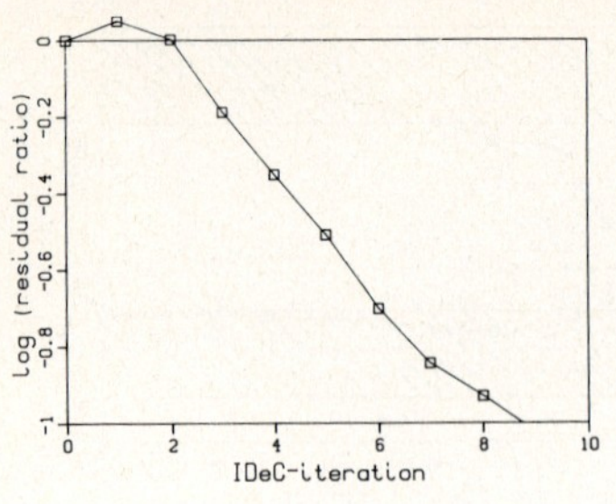

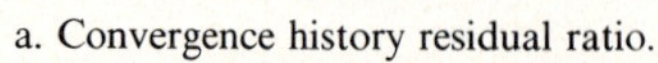

a. Convergence history residual ratio.

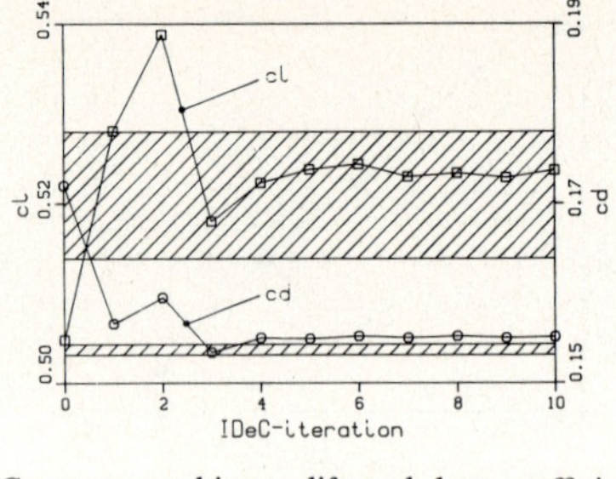

b. Convergence history lift and drag coefficient.

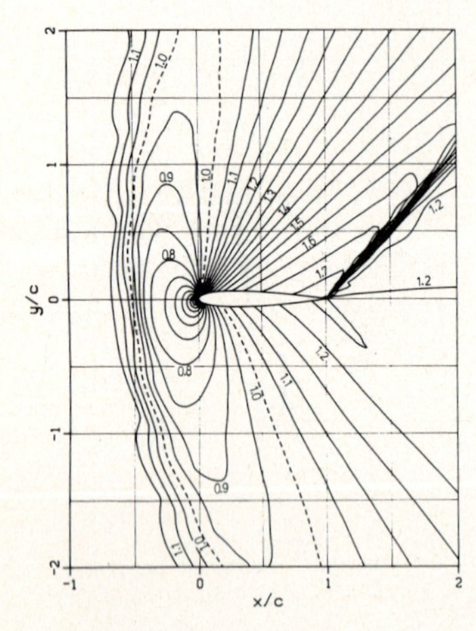

c. Mach number distributions; present result (left) and result Veuillot & Vuillot (right).

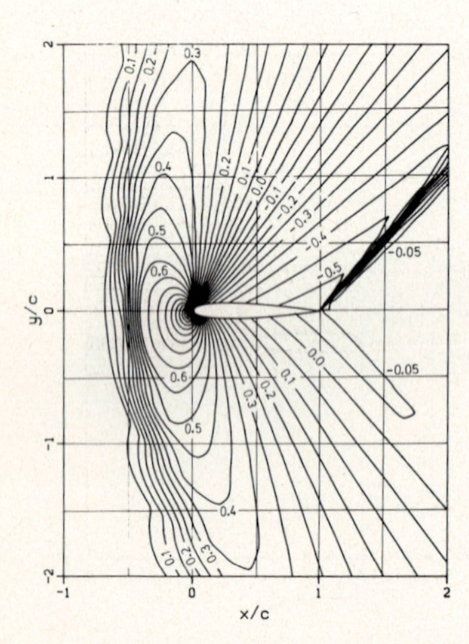

d. Present pressure distribution (c_p).

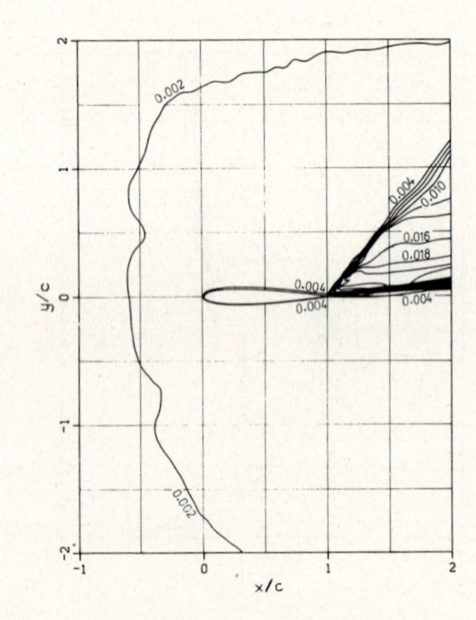

e. Present entropy distribution ($s/s_{\infty}-1$).

Fig. 4.4: Results for NACA0012-airfoil at $M_{\infty}=1.2$, $\alpha=7°$.

The NACA0012-bi-airfoil with propeller disk:
For this configuration we use a 128×48-grid as shown in fig. 4.5. The non-smoothness of the grid at the airfoil noses possibly leads to a solution of worse quality. In [10], where only a first-order accurate discretization was used, we already observed a spurious entropy rise along a kinked wall. However, to investigate the capabilities of a second-order discretization with respect to this non-smooth grid, and simply to avoid extensive grid generation efforts we prefer the present grid to a smooth grid in a multi-domain approach. The outer boundary of the grid is taken at an approximate distance from the configuration of 10 chord lengths. For the solution algorithm 5 IDeC-iterations and 4 coarser grids are taken.

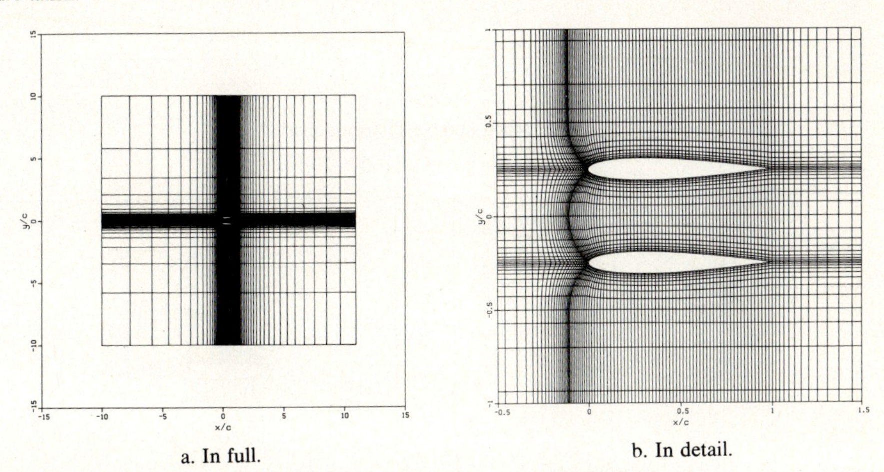

a. In full. b. In detail.

Fig. 4.5: 128×48-grid NACA0012-bi-airfoil with propeller disk.

We consider the configuration twice at $M_\infty = 0.5$, $\alpha = 2°$; namely with the propeller disk switched off and with the propeller disk switched on. In both cases we impose unperturbed conditions at the outer boundary ($u = M_\infty \cos\alpha$, $v = M_\infty \sin\alpha$ and $c = 1$ at inflow, and $p = 1$ at outflow). We assume the propeller disk to be located inside the volumes, i.e. we assume that the propeller does not coincide with any volume wall. In each volume $\Omega_{i,j}$ which is intersected by the propeller disk, a local line source of x-momentum and energy per unit of length and time ($(\delta_2)_{i,j}$ respectively $(\delta_4)_{i,j}$) are computed in the following way. We assume a sudden rise in the pressure:

$$p = (1+\delta) p_{i,j}, \tag{4.1}$$

with δ constant. Further, following [12,ch.3] we assume the flow through the propeller disk to be isentropic:

$$p \rho^{-\gamma} = p_{i,j} \rho_{i,j}^{-\gamma}. \tag{4.2}$$

With these two relations and with the steady jump relations

$$\begin{aligned}
&\rho u - \rho_{i,j} u_{i,j} = 0, \\
&(\rho u^2 + p) - (\rho_{i,j} u_{i,j}^2 + p_{i,j}) = (\delta_2)_{i,j}, \\
&\rho u v - \rho_{i,j} u_{i,j} v_{i,j} = 0, \\
&\{\frac{\gamma}{\gamma-1} p + \tfrac{1}{2}\rho(u^2 + v^2)\} u - \{\frac{\gamma}{\gamma-1} p_{i,j} + \tfrac{1}{2}\rho_{i,j}(u_{i,j}^2 + v_{i,j}^2)\} u_{i,j} = (\delta_4)_{i,j},
\end{aligned} \tag{4.3}$$

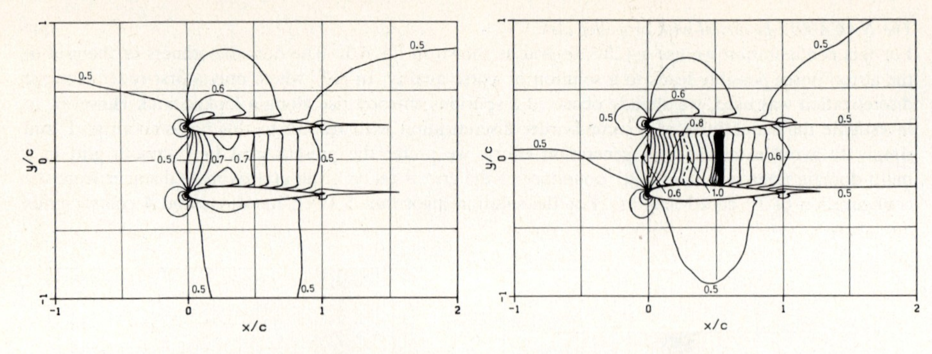

a. Mach number distributions.

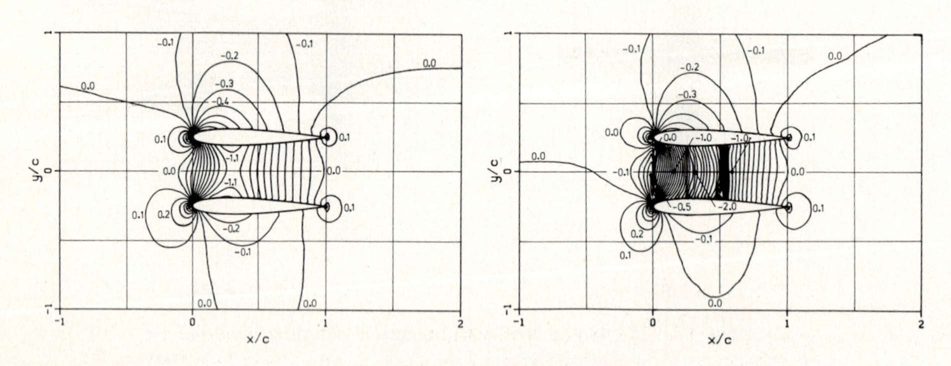

b. Pressure distributions (c_p).

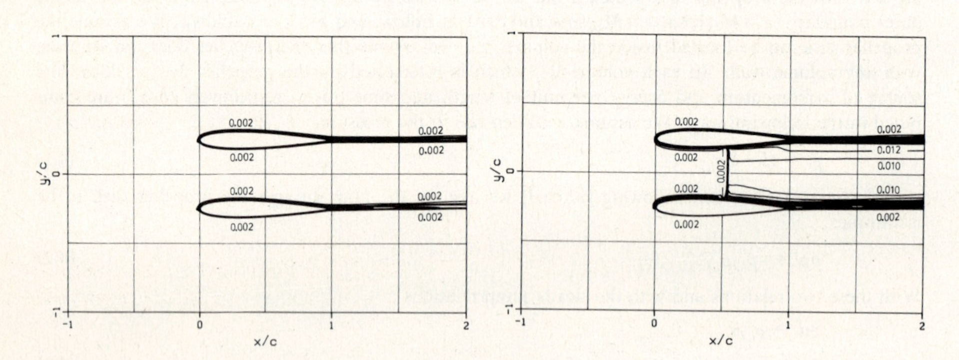

c. Entropy distributions ($s/s_\infty - 1$).

Fig. 4.6: Results for NACA0012-bi-airfoil with propeller disk, at $M_\infty = 0.5$, $\alpha = 2°$, propeller off (left), propeller on (right).

we have obtained a system of 6 equations with 6 unknowns $(\rho,u,v,p,(\delta_2)_{i,j}$ and $(\delta_4)_{i,j})$ from which $(\delta_2)_{i,j}$ and $(\delta_4)_{i,j}$ can be computed. (So in each volume $\Omega_{i,j}$ which is intersected by the propeller disk, the x-momentum and energy source are functions of δ and $q_{i,j}$ only.) In the IDeC-process(3.8) we take $r_h = r_h(q_h^{(n)})$, with $(r_h(q_h^{(n)}))_{i,j} = l_{i,j}(0,(\delta_2)_{i,j}^{(n)},0,(\delta_4)_{i,j}^{(n)})^T$, where $l_{i,j}$ is the length of the propeller part inside $\Omega_{i,j}$. The use of source terms which depend on the solution fits perfectly well in the IDeC-process.

For the case with working propeller we take $\delta=0.1$. For both cases (propeller on and off) we observe convergence to the solution of $F_h^2(q_h)=r_h(q_h)$. The values of lift and drag of the upper and lower airfoil (after the 5th IDeC-iteration) are given below.

Tab. 4.1: Lift and drag coefficients for NACA0012-bi-airfoil with propeller disk, at $M_\infty=0.5$, $\alpha=2°$.

	propeller disk off ($\delta=0$)	propeller disk on ($\delta=0.1$)
$c_{l_{upper}}$	-0.24	-0.89
$c_{l_{lower}}$	0.59	1.26
$c_{d_{upper}}$	-0.01	0.01
$c_{d_{lower}}$	0.02	0.06

Clearly visible is the large influence of the working propeller on the lift force acting on each of the two airfoils.

In fig. 4.6 we give the Mach number, pressure and entropy distribution as obtained for both cases. In the Mach number distribution for the case with working propeller we clearly observe the Mach number increase and shock wave, which have been introduced in the internal flow part (and, consistently, the Mach number decrease in the external flow part). The imposed sources are such that the flow has developed from fully subsonic into transonic. In the pressure distribution (fig. 4.6b) for the case with working propeller we observe that the assumed pressure jump appears indeed. Further, the corresponding entropy distribution (fig. 4.6c) shows the assumed isentropy through the propeller. For both cases (propeller on and off), the entropy which is spuriously generated at the airfoil noses and convected downstream, is an order of magnitude larger than for the single-airfoil. Cause of this is the non-smoothness of the grid at the noses of the bi-airfoil.

5. Conclusions

The Fromm scheme supplied with the Van Albada limiter yields second-order accurate solutions without spurious non-monotonicity and with sharp discontinuities. Comparison with the results of other investigators shows that for flows with discontinuities we obtain solutions of the same good quality on a grid which may be twice as course (in each coordinate direction).

For the computation of airfoil flows with the steady Euler equations, Iterative Defect Correction and nonlinear multigrid are found to be very efficient tools. It appears that it is sufficient to perform only a few IDeC-iterations; each implying only a single FAS-iteration.

An important property of the present method is that it is completely parameter-free; it needs no tuning of parameters.

Acknowledgements

The authors would like to thank J.W. Boerstoel and A. Kassies for providing O-type grids, P. Wesseling and P.W. Hemker for reading the manuscript, and H. Viviand and AGARD for permission to use reference results.

REFERENCES

1. G.D. VAN ALBADA, B. VAN LEER & W.W. ROBERTS (1982): *A Comparative Study of Computational Methods in Cosmic Gasdynamics.* Astron. Astrophys. 108, 76-84.
2. K. BÖHMER, P.W. HEMKER & H.J. STETTER (1984): *The Defect Correction Approach.* Computing, Suppl. 5, 1-32.
3. A. BRANDT (1982): *Guide to Multigrid Development.* Lecture Notes in Mathematics 960, Springer Verlag.
4. S.K. GODUNOV (1959): *Finite Difference Method for Numerical Computation of Discontinuous Solutions of the Equations of Fluid Dynamics* (in Russian, also Cornell Aeronautical Lab. Transl.). Math. Sbornik 47, 272-306.
5. W. HACKBUSCH (1985): *Multi-Grid Methods and Applications.* Springer Verlag.
6. A. HARTEN, P. LAX & B. VAN LEER (1983). *On Upstream Differencing and Godunov-type Schemes for Hyperbolic Conservation Laws.* SIAM Review 25, 35-61.
7. P.W. HEMKER (1985): *Defect Correction and Higher-Order Schemes for the Multigrid Solution of the Steady Euler Equations.* Report NM-R8523, Centre for Mathematics and Computer Science, Amsterdam. To appear in Proceedings 2nd European Multigrid Conference, Cologne, 1985. Lecture Notes in Mathematics, Springer Verlag.
8. P.W. HEMKER & B. KOREN (1986): *A Non-linear Multigrid Method for the Steady Euler Equations.* Report NM-R8621, Centre for Mathematics and Computer Science, Amsterdam. To appear in Proceedings GAMM-Workshop on The Numerical Simulation of Compressible Euler Flows, Rocquencourt, 1986. Vieweg Verlag Series Notes on Numerical Fluid Mechanics.
9. P.W. HEMKER & S.P. SPEKREIJSE (1985): *Multiple Grid and Osher's Scheme for the Efficient Solution of the Steady Euler Equations.* Report NM-R8507, Centre for Mathematics and Computer Science, Amsterdam. To appear in Appl. Num. Math., 1986.
10. B. KOREN (1986): *Euler Flow Solutions for a Transonic Windtunnel Section.* Report NM-R8601, Centre for Mathematics and Computer Science, Amsterdam.
11. B. KOREN (1986): *Evaluation of Second Order Schemes and Defect Correction for the Multigrid Computation of Airfoil Flows with the Steady Euler Equations.* Report NM-R8616, Centre for Mathematics and Computer Science, Amsterdam.
12. D. KÜCHEMANN (1978): *The Aerodynamic Design of Aircraft.* Pergamon Press.
13. B. VAN LEER (1982): *Flux-Vector Splitting for the Euler Equations.* Proceedings 8th International Conference on Numerical Methods in Fluid Dynamics, Aachen, 1982. Lecture Notes in Physics 170, Springer Verlag.
14. B. VAN LEER (1985): *Upwind-Difference Methods for Aerodynamic Problems governed by the Euler Equations.* Lectures in Applied Mathematics 22, AMS.
15. H.W. LIEPMANN & A. ROSHKO (1966): *Elements of Gasdynamics.* Wiley.
16. S. OSHER & F. SOLOMON (1982): *Upwind Difference Schemes for Hyperbolic Systems of Conservation Laws.* Math. Comp. 38, 339-374.
17. S. OSHER & S. CHAKRAVARTHY (1983): *Upwind Schemes and Boundary Conditions with Applications to Euler Equations in General Geometries.* J. Comp. Phys. 50, 447-481.
18. P.L. ROE (1981): *Approximate Riemann Solvers, Parameter Vectors, and Difference Schemes.* J. Comp. Phys. 43, 357-372.
19. S.P. SPEKREIJSE (1985): *Second-Order Accurate Upwind Solutions of the 2D Steady Euler Equations by the Use of a Defect Correction Method.* Report NM-R8520, Centre for Mathematics and Computer Science, Amsterdam. To appear in Proceedings 2nd European Multigrid Conference, Cologne, 1985. Lecture Notes in Mathematics, Springer Verlag.
20. S.P. SPEKREIJSE (1986): *Multigrid Solution of Monotone Second-Order Discretizations of Hyperbolic Conservation Laws.* Report NM-R8611, Centre for Mathematics and Computer Science, Amsterdam. To appear in Math. Comp.
21. S.P. SPEKREIJSE (1986): *A Comparison of Several Multigrid Methods for the Solution of Second-Order Upwind Discretizations of the Steady Euler Equations.* Report NM-R86xx, Centre for Mathematics and Computer Science, Amsterdam.
22. J.L. STEGER & R.F. WARMING (1981): *Flux-Vector Splitting of the Inviscid Gas Dynamic Equations with Applications to Finite-Difference Methods.* J. Comp. Phys. 40, 263-293.
23. P.K. SWEBY (1984): *High Resolution Schemes using Flux Limiters for Hyperbolic Conservation Laws.* SIAM J. Num. Anal. 21, 995-1011.
24. H. VIVIAND (1985): *Numerical Solutions of Two-dimensional Reference Test Cases.* In: Test Cases for Inviscid Flow Field Methods. H. YOSHIHARA, et al. (eds.). AGARD Advisory Report 211.

SOME EXPERIENCES WITH SPECTRAL METHODS

A.J. Renkema, R. Verstappen,
R.W. de Vries, P.J. Zandbergen,
Twente University of Technology,
P.O. Box 217, 7500 AE Enschede,
The Netherlands.

1. INTRODUCTION

Since the beginning of the seventies there is a growing interest in spectral methods as introduced by S.A. Orszag in 1970 [1]. Especially for problems with a high degree of continuity these methods can lead to very accurate results. In general problems are treated on simple geometries such as squares and boxes, either directly or by transformation of the region of interest.

For problems with periodic boundary conditions use is made of Fourier series, while for problems with Dirichlet or Neumann conditions use is made of Chebyshev-series. It can be shown that the convergence rate of Fourier series is at most $1/N^3$ where N is the number of modes taken into account while for Chebyshev series the convergence rate is faster than any power of N, the socalled exponential convergence. For an introduction to Chebyshev polynomials, the reader is referred to [2].

A drawback of spectral methods is that the involved matrices in general are full and that hence the inversion of the system of equations can be very time consuming when compared to modern methods for solving systems with sparse matrices.

This problem can be resolved in several ways. One way is to use symmetry and the special structure of the matrices involved, using a separation of variables technique. This has for instance been considered in [3] and [4].

A more general method which does not use the special properties of the matrices can be obtained by using a relaxation procedure such as Richardson relaxation. As is known the convergence rate of such a scheme is very poor if we try to smooth all the modes occurring in the Fourier representation of the error. As pointed out by A. Brandt [5] a relaxation scheme works efficiently on the high modes of the error and so he introduced the so called multi-grid method, which plays at present a dominant role in the application of finite difference- and finite element techniques.

In spectral methods functions are represented by Fourier or Chebyshev series, so it is quite natural that multigrid methods are introduced, especially in the so called pseudo-spectral method. Recently Zang et al. [6,7] and Brandt et al. [8] introduced the multigrid method in spectral theory, while in [9] applications are given for the Navier-Stokes equations.

In this paper we will give some of our own experiences with spectral methods, where the main emphasis will be on the application for the Navier-Stokes equations. The two different approaches selected above have both been used and although there are no final results, a number of remarks can be made which may be of some value for other people who want to use spectral methods.

When trying to solve the Navier-Stokes equations there is one phenomenon which has been the subject of many investigations and that is the proper description of the boundary-conditions.

For intance in two dimensions, when using the ψ,ω formulation there are no boundary conditions for ω, while when using the primitive variables formulation there are no conditions for the pressure. As has been shown by Kleiser [10] a so called influence function-technique may be used, but its application is certainly not without traps.

In seems quite clear that it would be very attractive to devise a technique which is capable of avoiding such very complicated methods. Some ideas on this matter will be formulated, when trying to solve two-dimensional Navier-Stokes problems in the ω-ψ formulation.

2. DIFFERENT FORMULATIONS OF THE NAVIER-STOKES PROBLEM

The Navier-Stokes equations for an incompressible fluid inside a domain B are written as

$$\frac{\partial \underline{u}}{\partial t} = \underline{u} \times \text{curl } \underline{u} - \text{grad } s + \nu \Delta \underline{u} \tag{2.1}$$

and

$$\text{div } \underline{u} = 0 \tag{2.2}$$

with $s = p + \frac{1}{2}(\underline{u} \cdot \underline{u})$ the stagnation pressure. On the edge ∂B of B appropriate boundary conditions are given, as for instance Dirichlet conditions for the velocity $\underline{u}$.

By taking the convergence of eq. (2.1) and applying eq. (2.2) there is obtained

$$\Delta s = \text{div } (\underline{u} \times \text{curl } \underline{u}) . \tag{2.3}$$

It can be readily shown that the system (2.1), (2.3) is equivalent with (2.1), (2.2) if there is required

$$\text{div } \underline{u} = 0 \qquad \text{for } t = 0 \tag{2.4}$$

$$\text{div } \underline{u} = 0 \qquad \text{on } \partial B \tag{2.5}$$

When solving the system (2.1)-(2.3) numerically we use a second order accurate discretization in time by using a Crank-Nicholson like technique.

$$\underline{u}^{n+1} - \underline{u}^{n} = \Delta t \{ \tfrac{3}{2} \underline{M}^{n} - \tfrac{1}{2} \underline{M}^{n-1} - \text{grad } s^{n+\frac{1}{2}} + \tfrac{\nu}{2} (\Delta \underline{u}^{n+1} + \Delta \underline{u}^{n}) \} \tag{2.6}$$

$s^{n+\frac{1}{2}}$ is determined from

$$\Delta s^{n+\frac{1}{2}} = \text{div} \{ \tfrac{3}{2} \underline{M}^{n} - \tfrac{1}{2} \underline{M}^{n-1} \} \tag{2.7}$$

where $\underline{M} = \underline{u} \times \text{curl } \underline{u}$.

The final system to be solved can be written as

$$\Delta \underline{u}^{n+1} - \lambda \underline{u}^{n+1} = \text{grad } q^{n+\frac{1}{2}} - \underline{r}^{n}, \tag{2.8}$$

$$\Delta q = \text{div } \underline{r}^n \tag{2.9}$$

with $\lambda = \frac{2}{\nu \Delta t}$; $\underline{r}^n = \frac{2}{\nu}[\frac{3}{2}\underline{M}^n - \frac{1}{2}\underline{M}^{n-1} - \frac{1}{\Delta t}\underline{u}^n] + \Delta \underline{u}^n$, $q^{n+\frac{1}{2}} = \frac{2}{\nu} s^{n+\frac{1}{2}}$.

As will be evident from equations (2.8) and (2.9) the solution of this time-dependent Navier-Stokes problem is obtained by solving two Helmholtz equations and one Laplace equation in each time step. The boundary conditions that we will consider are (2.5) and $\underline{u}$ is prescribed on ∂B.

To obtain the ω-ψ formulation for two-dimensional flows we pose

$$\underline{u} = \text{curl } \psi \underline{k} \tag{2.10}$$

and

$$\text{curl } \underline{u} = -\omega \underline{k} . \tag{2.11}$$

Taking the curl of eq. (2.10) and of eq. (2.1) leads to the system

$$\Delta \psi = \omega \tag{2.12}$$

$$\frac{\partial \omega}{\partial t} + \text{curl } \psi \underline{k} \cdot \text{grad } \omega = \nu \Delta \omega . \tag{2.13}$$

If Dirichlet conditions for $\underline{u}$ are given, this means that ψ and $\frac{\partial \psi}{\partial n}$ are given along ∂B.

In the sequel we will first consider the solution of the time dependent problem, thereafter we will make a few remarks on the ω-ψ problem.

3. SOLUTION OF THE TIME DEPENDENT PROBLEM IN PRIMITIVE VARIABLES

In order to solve the time dependent problem as formulated above, we have to solve for each time step the Helmholtz equations (2.8) and the Laplace equation (2.9). For the region B we consider the square $\{x,y \mid -1 \leq x,y \leq 1\}$ and we will assume that the boundary conditions are: $\underline{u}$ is given along ∂B, together with eq. (2.5). In fact we are then missing boundary conditions for q and we will use Kleiser's method [10] to treat this problem. For this purpose we split the problem in a particular part and a homogeneous part.

$$\underline{u} = \underline{u}_p + \underline{u}_h \text{ , } q = q_p + q_h . \tag{3.1}$$

The particular solution is determined from eqs. (2.8) and (2.9) by posing

$$\begin{aligned} \underline{u}_p &= \underline{u} \text{ prescribed along } \partial B \\ q_p &= 0 \qquad \text{along } \partial B. \end{aligned} \tag{3.2}$$

The homogeneous solution is determined as follows. We write

$$q_h = \sum \alpha_i q_i \qquad \underline{u}_h = \sum \alpha_i \underline{u}_i \tag{3.3}$$

and solve

$$\Delta q_i = 0 \qquad \text{with} \qquad \begin{array}{l} q_i = 1 \text{ in edge point } i \\ q_i = 0 \text{ in edge point } i \neq j \end{array} \tag{3.4}$$

$$\underline{\Delta} \underline{u}_i - \lambda \underline{u}_i = \text{grad } q_i \quad \text{with} \quad \underline{u}_i = 0 \quad \text{on} \quad \partial B. \tag{3.5}$$

The influence coefficients α_i are then determined from the equation

$$\sum_i \alpha_i (\text{div } \underline{u}_i)_j = - (\text{div } \underline{u}_p)_j \, . \tag{3.6}$$

We will describe the actual solution by explaining the way in which the Helmholtz equation is solved, using a spectral method combined with a separation of variables technique.
The solution is represented as

$$u = \sum_{n=0}^{N} \sum_{m=0}^{N} u_{nm} \, T_n(x) T_m(y) \, . \tag{3.7}$$

We use the so called τ method which means that 4N quantities $U_{N-i,m}$, $u_{n,N-i}$; $i = 0,1$ are eliminated by using the boundary conditions and that the differential equations have to be fulfilled at a number of internal collocation points $x_i = \cos \frac{i\pi}{N}$, $i = 1\text{-}N\text{-}1$.
This leads to the following equation

$$AU + U^{0,2} - \lambda U = F - R^* \tag{3.8}$$

where A is a $(N-1)\times(N-1)$ matrix representing the $\frac{\partial^2}{\partial x2}$ operator and $U^{0,2}$ is the $\frac{\partial^2}{\partial y2}$ oparator. F is the source term and R* is the result of the elimination procedure and hence contains the boundary conditions.
The separation of variables technique leads to the diagonalisation of A by introducing

$$A = CVC^{-1} \qquad U = CZ \qquad G = C^{-1}(F-R^*) \tag{3.9}$$

which gives

$$VZ + Z^{0,2} - \lambda Z = G \, . \tag{3.10}$$

As has been shown in [3] and [4] this equation can be solved very efficiently once the eigenvalues of A are known. It will be immediately clear that every problem on an interval can be splitted into even and uneven solutions, and that hence a transformation may be found which reflects this property. This indeed is the case. The eigenvalues can be found from two matrices of the rank $\frac{N}{2}$ and $\frac{N}{2} - 1$, from which moreover the eigenvectors can be calculated easily. In this way a large reduction in computer time can be obtained.
One of the drawbacks of this method is that the accuracy of the eigenvalues is decisive for the accuracy of the final results. It turns out that even when performing the diagonalisation in double precision on a CDC computer (28 decimal places) the eigenvalues for N = 64 cannot be calculated accurately enough.

So far we have not spoken about the determination of the influence coefficients α_i according to eq. (3.6). It should be observed that this leads to a 4N×4N matrix which will be full.

However, as will be immediately clear there is 8 fold symmetry, since the solution will be symmetric along the lines $x = 0$, $y = 0$ and $x = \pm y$. There is however another problem and that is that the influence matrix is five fold singular. This is a direct consequence of the τ method. The differential equations will be solved in $(N-1)\times(N-1)$ polynomials, the remaining ones are found from the boundary conditions. For the homogeneous problem as described in (3.3), (3.4) and (3.5) this means that the highest coefficients of q_h are not determined. So we find that

$$q_h = \sum_{i=1}^{5} q_{0,0}T_0(x)T_0(y) + q_{N-1,N-1}T_{N-1}(x)T_{N-1}(y)$$

$$+ q_{N-1,N}T_{N-1}(x)T_N(y) + q_{N,N-1}T_{Nm}(x)T_{N-1}(y)$$

$$+ q_{N,N}T_N(x)T_N(y)$$

$$u_h = v_h = 0 \tag{3.11}$$

is a solution of the homogeneous problem.
In order to make the system non-singular, it is possible to choose the four highest coefficients of q equal to zero.

The method has been tested by applying it to the model problem (ref. 11)

$$u(x,y,t) = - \cos \lambda x \sin \lambda y \; e^{-2\lambda^2 t/Re}$$

$$v(x,y,t) = \sin \lambda x \cos \lambda y \; e^{-2\lambda^2 t/Re} \tag{3.12}$$

$$p(x,y,t) = - \frac{1}{4} \{\cos(2\lambda x) + \cos(2\lambda y)\} e^{-4\lambda^2 t/Re}$$

with $\lambda = \frac{\pi}{2}$ and $Re = 50$.

For $t = 1$ the results agreed in 8 digits for $\underline{u}$ and p when applying $N = 32$.

There can however be difficulties when applying the τ method in obtaining and maintaining the divergence free condition. We feel this is so important that we will look into this matter a little further.

4. SOME REMARKS ABOUT THE KLEISER METHOD

We start again with the particular solution as given by eqs. (3.1) and (3.2). We hence can calculate

$$\text{div } \underline{u}_p \, .$$

Now considering eqs. (3.3), (3.4) and (3.5) we remark that we can write

$$\underline{u}_h = \text{grad } \phi + \text{rot } \psi \underline{k} - \frac{1}{\lambda} \text{grad } q_h \, . \tag{4.1}$$

When taking the divergence of $\underline{u}_h$ we find

$$\text{div } \underline{u}_h = \Delta\phi - \frac{1}{\lambda} \Delta q_h = \Delta\phi. \tag{4.2}$$

On the other hand we know that there should hold

$$\Delta \operatorname{div} \underline{u} - \lambda \operatorname{div} \underline{u} = 0$$

with $\operatorname{div} \underline{u} = 0$ along ∂B

and hence $\operatorname{div} \underline{u} \equiv 0$ in $B \cup \partial B$

which means that $\operatorname{div} \underline{u}_p = - \operatorname{div} \underline{u}_h$. (4.3)

Substituting (4.1) into (3.5) leads to

$$\operatorname{grad} (\Delta\phi - \lambda\phi) + \operatorname{rot} (\Delta\psi - \lambda\psi)\underline{k} = 0 . \tag{4.4}$$

If we now take

$$\phi = - \frac{1}{\lambda} \operatorname{div} \underline{u}_p \tag{4.5}$$

we find that $\underline{u}_h$ finally can be written as

$$\underline{u}_h = + \frac{1}{\lambda} \operatorname{grad} \operatorname{div} \underline{u}_p - \frac{1}{\lambda} \operatorname{grad} q_h + \operatorname{rot} \psi\underline{k} \text{ along } \partial B \tag{4.6}$$

which gives 8 boundary conditions along the faces ∂B of B for the 8 functions q_n and ψ along the different faces. In this way we solve only for q_i and ψ_i, however instead of the unknown coefficients α_i of q_i, we also have the unknown coefficients β_i of ψ_i. Hence the influence matrix becomes twice as large.

It may seem that we in this way have evaded the problems with $\operatorname{div} \underline{u}$ since analytically it is quite clear that this should be zero everywhere. However, when, using the spectral method starting with the representation (3.7), we loose the highest powers of x^n and y^n, when calculating (4.6), which prevents $\operatorname{div} \underline{u}_h$ from being zero in all the $(N+1)\times(N+1)$ spectral coefficients as it should be according to (4.3).

The only way out of this dilemma is to enforce the solution of (4.3) by solving u, v and q for different numbers of coefficients

$$u = \sum_0^{N+1} \sum_0^{N} N_{nm} T_n(x) T_m(y)$$

$$v = \sum_0^{N} \sum_0^{N+1} v_{nm} T_n(x) T_m(y)$$

while $$q = \sum_0^{N} \sum_0^{N} q_{nm} T_n(x) T_m(y)$$

and using the original system (3.3)-(3.5).

A drawback is that when using this, we can no longer take $N = 2^k$, and hence can no longer use Fast Fourier Transforms.

The total amount of operations is estimated for ℓ time steps as

$$(\tfrac{3}{2} N + 23)N^3 + 9\ell N^3 .$$

Although this may seem a large number, there are in the literature several examples of computations which when performed with finite differences would have taken grids of say 1000×1000 for a problem which can be solved by 32×32 polynomials. This means that for these problems the complexity of the method is still very moderate.

There are however of course a number of disadvantages which can be mentioned. As mentioned in ref. 9, variable viscosity prevents the velocity and pressure equations from uncoupling and thus the use of influence matrix methods. The matrix diagonalisation technique is not practical when A depends on time.

This has led to the development of spectral multigrid methods, and we will give our own experiences with this method in the following section.

5. INTRODUCTION TO SPECTRAL MULTIGRID METHODS

To outline the essentials of the method we will first consider a periodic one-dimensional problem for

$$\frac{d^2u}{dx^2} = f \,. \tag{5.1}$$

For a periodic problem the Fourier Transform is

$$\hat{u}_n = \frac{1}{N}\sum_{j=0}^{N-1} u_j e^{-2\pi ij\, n/N} \qquad p = -\frac{N}{2}, -\frac{N}{2}+1 \dots \frac{N}{2}-1$$

and

$$u_j = \sum_{p=-\frac{N}{2}}^{\frac{N}{2}-1} \hat{u}_p e^{2\pi ijp/N} \qquad j = 0,1,\cdots,N-1$$

where

$$x_j = \frac{2\pi j}{N} \,.$$

Equation (5.1) can then be written as

$$C^{-1}\, D\, C\, U = F \tag{5.2}$$

where C, C^{-1} are matrices with elements

$$C_{pj} = \frac{1}{N} e^{-2\pi ipj/N}$$

$$C^{-1}_{jp} = e^{2\pi ijp/N}$$

$$D_{pq} = -p^2 \delta p_q$$

$$D_{pp} = 0 \quad \text{for } |p| = N/2 \,.$$

If equation (5.2) is solved iteratively by Richardson relaxation on a single grid we have

$$v^{(k+1)} = v^{(k)} - \omega\left(F - L_{sp} v^{(k)}\right). \tag{5.3}$$

The spectral radius of the iteration matrix $I + \omega L$ is $\rho = 1 + \omega\lambda_{max}$ where λ_{max} is the largest eigenvalue of $C^{-1}DC = L_{sp}$.

The optimal choice for ω is

$$\omega_{opt} = \frac{2}{\lambda_{min} + \lambda_{max}} \qquad \text{and so we find}$$

$$\rho = \frac{\lambda_{max} - \lambda_{min}}{\lambda_{max} + \lambda_{min}} \sim 1 - \frac{8}{N^2}$$

in the case we are considering. For N increasing, ρ tends to unity and hence convergence will be slow.

If we however base the smoothing factor ρ on the middle eigenvalue of $C^{-1}DC$ we find $\rho = \frac{3}{5}$. This means that now the higher modes are relaxed much faster, but the lower modes are unaffected by the relaxation proces. But the lower modes can be smoothed on a coarser grid. So we arrive at a multigrid method by performing some relaxiations on a fine grid and then changing to a coarser grid and if necessary repeating this process. Restriction and prolongation can be found by using trigonometric interpolation but we will at once proceed to the Dirichlet problem for Poisson and Helmholtz equations.

6. MULTIGRID DIRICHLET PROBLEMS

In this section we consider the problem

$$Lu = f \quad \text{on the square} \cdot -1 \leq x,y \leq 1 \tag{6.1}$$

with $L = \Delta$ or $L = \Delta - \lambda$.

The method which will be used is a pseudo-spectral method which essentially works in the fysical space. The collocation points are

$$x_i = \cos \frac{\pi i}{N} \quad \text{and} \quad y_j = \cos \frac{\pi j}{N} \quad i,j = 1,2,\ldots,N-1.$$

Since in this case the eigenvalues are much larger than in the Fourier case, we cannot use (5.3) directly, but a pseudo inverse H^{-1} of the matrix L_{sp} has to be used as preconditioning matrix.

$$v^{(k+1)} = v^{(k)} + \omega H^{-1} (L_{sp} v^{(k)} - F). \tag{6.2}$$

In our own experiments we have used the full LU decomposition of the finite difference matrix for the Laplace operator a preconditioning matrix.

In refs. [6] and [7] incomplete decomposition of the finite difference matrix is used, however as has been observed in ref. [12] conjugate gradient methods which are widely in use for finite difference or finite element approximations seem to require further analysis. This is in agreement with our own findings, for the ICCG method of ref. [13] showed fast convergence for equidistant collocation points, but no convergense for the cosine Chebyshev distribution.

The distribution and prolongation operators are given as follows

$$q_{N/2}(x_{2\ell}) = \sum_{r=0}^{N/2} \tilde{c}_r b_r T_r(x_{2\ell}) \tag{6.3a}$$

with

$$b_r = \frac{2}{N} \sum_{k=0}^{N} \bar{c}_k \, u(x_k) T_r(x_k) \qquad (6.3b)$$

and

$$\tilde{c}_0 = \bar{c}_0 = \tilde{c}_{N/2} = \bar{c}_N = \frac{1}{2} \quad \text{otherwise } \tilde{c}_r = \bar{c}_r = 1. \qquad (6.3c)$$

This is in effect taking only half of the coefficients from the fine grid to evaluate the values on the coarser grid, and to reduce the coefficient $b_{N/2}$ by a factor a half.

Inserting (6.3b) into (6.3c) we find for the restriction operator

$$R_{\ell k} = \frac{2}{N} \bar{c}_k \sum_{r=0}^{N/2} \tilde{c}_r \cos \frac{k\pi r}{N} \cos \frac{2\ell \pi r}{N} \, . \qquad (6.4)$$

In the same way the prolongation operator is given by

$$P_{kj} = \frac{4}{N} \tilde{c}_k \sum_{r=0}^{N/2} \tilde{c}_r \cos \frac{2k\pi r}{N} \cos \frac{j\pi r}{N} \qquad (6.5)$$

These operators are not adjoint, hence the equation

$$(Pu_{k-1}, v_k)_k = (u_{k-1}, Rv_k)_{k-1}$$

is not fulfilled, but the difference is only a factor of 2. This can be enforced by posing $\tilde{c}_k = \bar{c}_j$.

The two-dimensional problem can be treated as a repeated calculation in one dimension, using finite transforms and Chenshuv reduction [ref. 14]. The Dirichlet boundary conditions have been implemented by posing

$$v_b^{(k+1)} = v_b^{(k)} + \omega\left(v_b(\text{given}) - v_b^{(k)}\text{spectral}\right) . \qquad (6.6)$$

It turns out that this works quite well for Poisson and Helmholtz equations.

To get a good impression of the values of ω which should be applied, the eigenvalues of $H'L_{sp}$ were first calculated for $N = 4$ and $N = 8$. It turned out that the highest eigenvalues are not increasing very much with increasing N, while H^{-1} is so good that many eigenvalues are clustering around 1, as is indicated in the following table.

Eigenvalues of $A^{-1} L$

N = M	λ_{max}	λ_{min}
16	4.6	1.0
8	4.3	1.0
4	3.5	1.0

As an example we give the case of a Poisson equation for which the exact solution is

$$u = \sin(\pi \cos x + \pi/4) \sin(\pi \cos y + \pi/4) \, .$$

The right-hand side of this equation is given in fig. 1, while the solution is presented in fig. 2.

In fig. 3 we give the convergence history of the multigrid calculations. In this case we win an order of magnitude each time we have performed one multigrid cycle. In this example on each grid we relax 6 times. As can be seen we obtain machine accuracy when calculating this example with N = 16 for the finest grid.

The same results can be obtained when calculating the solution of Helmholtz equations. In this way we have the basic material to solve the equations (2.8) and (2.9) of the Navier-Stokes problem in the primitive variable formulation. This has so far not been implemented, but we will outline our ideas about the lines of further investigations below.

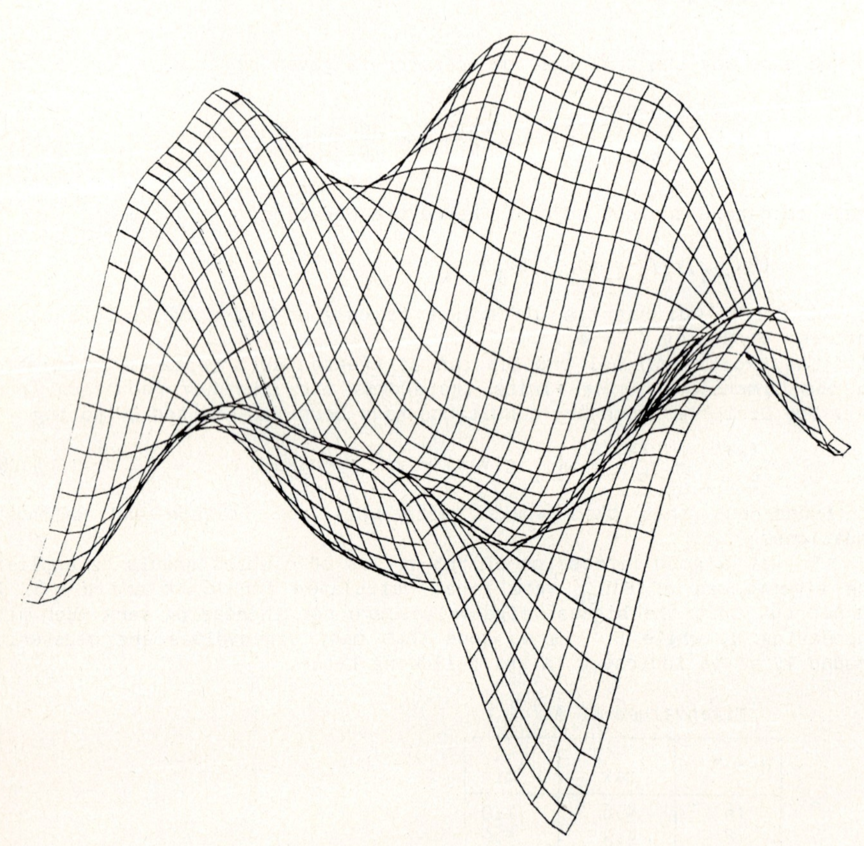

fig. 1: The right-hand side of the model problem

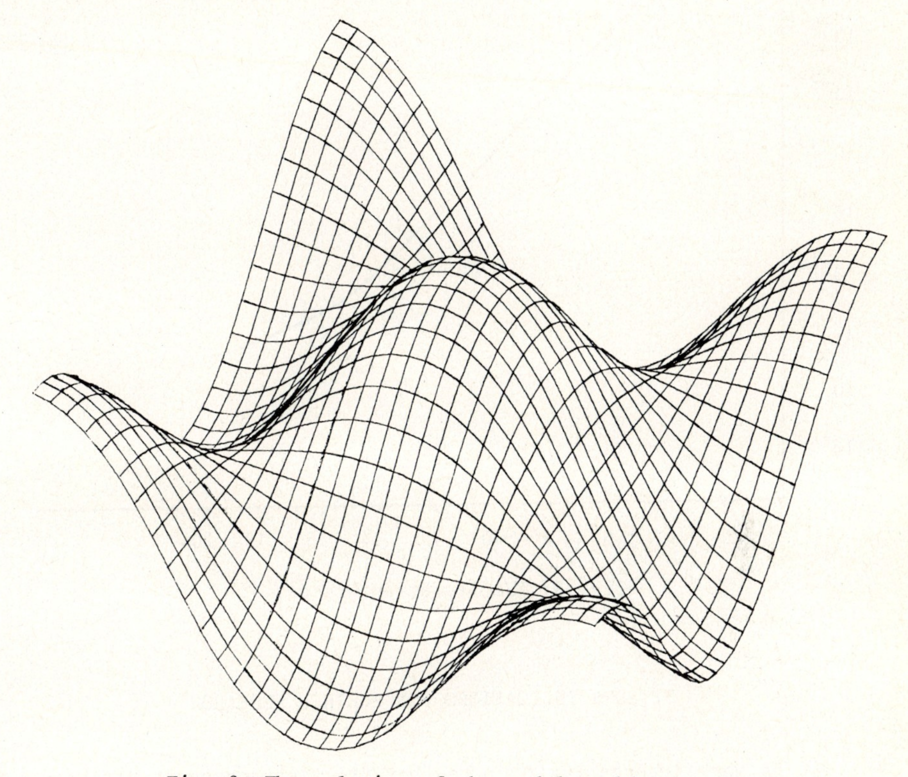

Fig. 2: The solution of the model problem

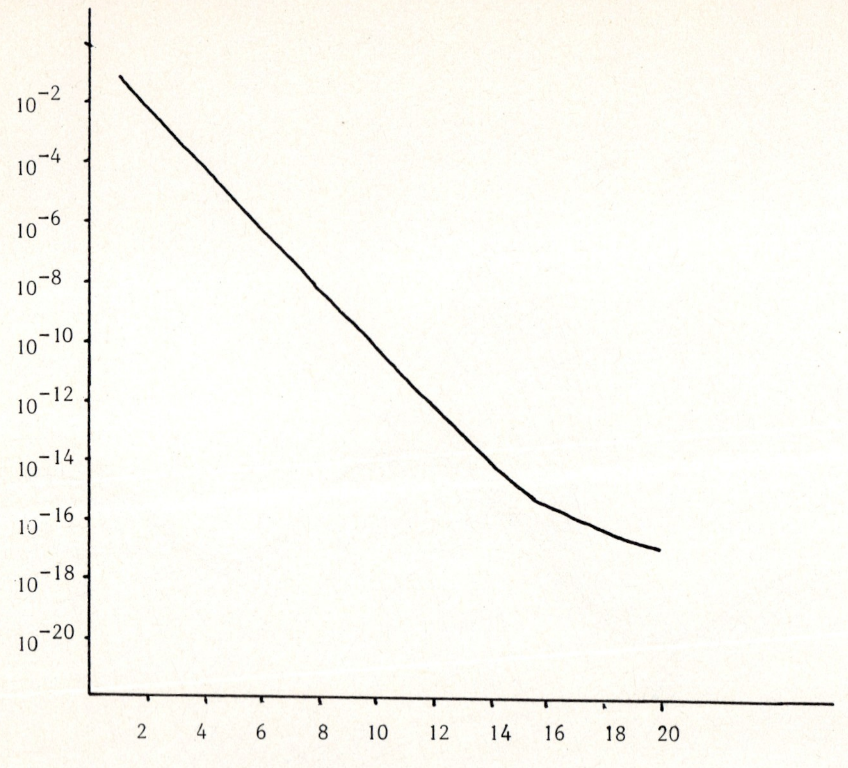

fig. 3: The iteration history

7. SOME INDICATIONS FOR FURTHER RESEARCH

It is very interesting to note that the time-dependent Navier-Stokes problem in the ω-ψ formulation when using the same time integration as in the $\underline{u}$, q formulation, leads to the following system of equations, where use has been made of eqs. (2.12) and (2.13).

$$\Delta\omega^{n+1} - \lambda\omega^{n+1} = \tilde{r}^{n} \tag{7.1}$$

$$\Delta\psi^{n+1} = \omega^{n+1} \tag{7.2}$$

and where $\tilde{r}^{n} = \frac{2}{\nu}\,[\frac{3}{2}\,N^{n} - \frac{1}{2}\,N^{n-1} - \frac{1}{\Delta t}\,\omega^{n}] + \Delta\omega^{n}$ and

$$N = \text{curl } \psi\underline{k}\cdot\text{grad } \omega. \tag{7.3}$$

Comparing (7.1), (7.2) with (2.8), (2.9) one immediately observes the big similarity between the two formulations. However this time the missing boundary condition is for ω, whereas ψ has one too much. In principle one could solve this problem by posing a number of Dirichlet problems for ω and then determine the unknown coefficients such that the two boundary conditions for ψ can be fulfilled.

Another way is to solve the biharmonie system

$$\Delta\Delta\psi^{n+1} - \lambda\Delta\psi^{n+1} = \tilde{r}^{n+1}$$

with boundary conditions for ψ and $\frac{\partial\psi}{\partial n}$.

A third way is to observe that ω along the boundary can be directly expressed in the unknown quantity $\frac{\partial^2\psi}{\partial n^2}$. F.i. we know that

$$\left(\frac{\partial^2\psi}{\partial y^2}\right)_{\partial B} = \omega_{\partial B} - \left(\frac{\partial^2\psi}{\partial x^2}\right) \quad \text{along} \quad y = 1 \, . \tag{7.4}$$

Now $\frac{\partial^2\psi}{\partial y^2}$ can be directly expressed in the values of ψ and $\frac{\partial x}{\partial y}$ on ∂B, and this means that in this way a direct connection can be obtained with the given values for ψ and $\frac{\partial\psi}{\partial y}$. In this way we can iteratively solve for the true solution.

This is directly connected with the problem of solving a mixed boundary value problem as for instance the problem with along two edges of the square given values of ψ and along the two other edges $\frac{\partial\psi}{\partial n}$.

For the one-dimensional case the above developed principle works, and as experience has shown application to the two-dimensional problem is straight-forward.

In the near future we hope to give results for these problems solved along the lines as sketched above.

REFERENCES

1. Gottlieb, D; Orszag, S.A. Numerical Analysis of spectral methods; theory and applications. Regional Conference Series in Applied Mathematics, Siam 1977.

2. Fox, L; Parker, I.B. Chebyshev Polynomials in Numerical Analysis Oxford University Press, London 1968.

3. Verstappen, R. Applications of the spectral method for the calculation of a two-dimensional incompressible flow. (In Dutch) Master Thesis, University Twente, Oct. '85.

4. Verstappen, R; Ten Thije Boonkamp, J.; De Vries, R; Zandbergen, P.J. Solution of the Navier-Stokes equation using an efficient spectral method. To appear in: Proceedings of the 10e International Conference on numerical methods in fluid dynamics.

5. Brandt, A. Multi-Level Adaptive Solutions to Boundary-Value Problems Mathematics of Computation, Volume 31, number 138, April 1977, p. 333-390.

6. Zang, T.A.; Wong, Y.S.; Hussaini, M.Y. — Spectral Multigrid methods for Elliptic Equations, Journ. of. Comp. Phys. 48, 1982 p 458-501.

7. Zang, T.A.; Wong, Y.S.; Hussaini, M.Y. — Spectral Methods for Elliptic Equations II. Journ. of Comp. Phys. 54, 1984, p 489-507.

8. Brandt, A; Fulton S.R.; Taylor, G.D. — Improved Spectral Multigrid Methods for Periodic Elliptic Problems. Journ. of Comp. Phys. 58, 1985, p. 96-112.

9. Malik, M.R.; Zang. T.A.; Hussaini, M.Y. — A Spectral Collocation Method for the Navier-Stokes Equations. Journ. of Comp. Phys. GI 1983, p. 64-88.

10. Kleiser, L; Schumann, U; — Spectral simulation of the Lamion turbulent transition process in plane Poiseuille flow. Proc. of ICASE symp. on Spectral Methods SIAM CBMS, 1983.

11. Taylor, G.I.; Green A.E. — Mechanism of the production of small eddies from large ones. Proceedings R. Soc. London, Ser. A 158, 1937, p. 499-521.

12. Canuto, C; Quateroni, A.; — Preconditioned Minimal Residual Methods for the iterative solution of systems of linear equations Journ. of Comp. Phys. 26, 1978 p. 43-65.

13. Kershaw, D.S. — The incomplete Cholesky Conjugate gradient method for the iterative solution of systems of linear equation Journal of Comp. Physics 26, 1978 p. 43-65.

WEAKLY-REFLECTIVE BOUNDARY CONDITIONS

FOR SHALLOW WATER EQUATIONS

G.K. Verboom[1] and A. Segal[2]

1) Delft Hydraulics, Estuaries and Seas Division, Rotterdamseweg 185, 2600 MH Delft, Holland,
2) Delft University of Technology, Department of Mathematics and Informatics, Julianalaan 132, 2628 BL Delft, Holland

SUMMARY

In this paper weakly-reflective boundary conditions are presented for the one- and two-dimensional shallow water equations. After a short discussion on the necessity of this type of boundary conditions the mathematical formulation of the problem is given together with the essential aspects of the derivation. The numerical implementation of these conditions in finite element and finite difference codes is considered and results are given for some applications.

INTRODUCTION

The flow in rivers, estuaries, and coastal seas can often adequately be described by the one- and two-dimensional shallow water equations. These equations describe a flow that is nearly horizontal and homogeneous in the vertical direction. Averaging of the Navier-Stokes equations over the turbulent fluctuations and the vertical direction may give rise to viscous terms. If these viscous terms are neglected the equations are known to be strictly hyperbolic [1]. For the unique solution of the shallow water equations initial and boundary conditions must be specified. In most practical applications at least part of the boundaries are purely artificial, defined to limit the domain in which a numerical solution is to be obtained. In nature, waves or more generally disturbances can freely cross these - non-existing - boundaries, but in a numerical solution this property must be included explicitely in the formulation of the boundary conditions. In practice this can only be realized to some extend i.e. the actual boundary conditions are not strictly non-reflective but only weakly-reflective to some order of approximation.
In engineering applications boundary conditions are taken from measurements, where the boundaries themselves are chosen outside the area of influence of the actions under study. With weakly-reflective boundary conditions the computational area can be decreased to within this area. In addition, weakly-reflective boundary conditions are very effective to get rid of initial disturbances, thereby decreasing the transient very substantial.

In this paper some of the major aspects are discussed of the derivation and of the numerical implementation of weakly-reflective boundary conditions for the two-dimensional shallow water equations. Results are given for both a finite element and a finite difference implementation.

MATHEMATICAL FORMULATION

THE DIFFERENTIAL EQUATIONS

The two-dimensional shallow water equations in primitive variables read

$$\vec{w}_t + A\vec{w}_x + B\vec{w}_y + C\vec{w} = \vec{0}, \tag{1}$$

with

$$\vec{w} = (u,v,\zeta)^T, \tag{2}$$

and

$$A = \begin{pmatrix} u & o & g \\ o & u & o \\ h+\zeta & o & u \end{pmatrix} ; \qquad B = \begin{pmatrix} v & o & o \\ o & v & g \\ o & h+\zeta & v \end{pmatrix} ,$$

$$C = \begin{pmatrix} \lambda & -f & o \\ f & \lambda & o \\ o & o & o \end{pmatrix} ,$$

where u and v are the x- and y-component of the depth integrated velocity vector, ζ is the free surface elevation above a reference plane, g is the gravitational acceleration, h is the water depth below a reference plane, f is the Coriolis parameter, λ is the bottom friction parameter, and x, y, t are the space coordinates and time, respectively. The bottom friction parameter λ is generally given by

$$\lambda = C_f \frac{\sqrt{u^2 + v^2}}{h + \zeta} , \tag{3}$$

where C_f is a friction factor.
In system (1) viscous terms and external forces are neglected. System (1) is a strictly hyperbolic system i.e. A en B have a complet set of eigenvectors and all eigenvalues are real. This is a very convenient, though not strictly required, property of the system when deriving weakly-reflective boundary conditions. When viscous terms are taken into account system (1) becomes a so-called incomplete parabolic system [2]. Weakly-reflective boundary conditions then only apply to the hyperbolic part of the equations. Later on a short discussion will be given on this matter.
Neglecting external forces in system (1) is not essential as they can easily be accounted for in the boundary conditions.
For the analysis in the following section it is advantageous to start from symmetrical matrixes A and B, at least one of which is a diagonal matrix. As system (1) is a quasi-linear strictly hyperbolic system A and B can be symmetrized (but not diagonalized) simultaneously, for instance with the following transformations

$$\psi = 2\sqrt{g\,(h+\zeta)}, \tag{4}$$

and

$$\vec{v} = V\,(u,v,\psi)^T, \tag{5}$$

where V is given by

$$V = \frac{1}{\sqrt{2}} \begin{pmatrix} 1 & 1 & 0 & 1 \\ (0 & \sqrt{2} & 0). \\ 1 & 0 & 1 \end{pmatrix} \tag{6}$$

In terms of the new variables $\vec{v}$ the system of equations (1) reads

$$\vec{v}_t + \tilde{A}\,\vec{v}_x + \tilde{B}\,\vec{v}_y + \tilde{C}\,\vec{v} = 0, \tag{7}$$

with

$$\vec{v} = \left(\frac{1}{\sqrt{2}}(u+\psi),\ v,\ \frac{1}{\sqrt{2}}(u-\psi)\right)^T, \tag{8}$$

$$\tilde{A} = \begin{pmatrix} u+\tfrac{1}{2}\psi & o & o \\ o & u & o \\ o & o & u-\tfrac{1}{2}\psi \end{pmatrix} \; ; \qquad \tilde{B} = \begin{pmatrix} v & \psi/2\sqrt{2} & o \\ \psi/2\sqrt{2} & v & -\psi/2\sqrt{2} \\ o & -\psi/2\sqrt{2} & v \end{pmatrix},$$

$$\tilde{C} = \begin{pmatrix} \lambda/2 & -f/\sqrt{2} & \lambda/2 \\ f/\sqrt{2} & \lambda & f/\sqrt{2} \\ \lambda/2 & -f/\sqrt{2} & \lambda/2 \end{pmatrix}.$$

WEAKLY-REFLECTIVE BOUNDARY CONDITIONS, SKETCH OF DERIVATION

The general solutions of the systems (1) and (7) contain progressive waves which, even in the case of vanishing C, i.e. no bottom friction and Coriolis force, are coupled. If these waves were uncoupled one could prescribe at a boundary the incoming wave and leave the outgoing wave undisturbed. Such boundary conditions are called non-reflective. The progressive waves would only be uncoupled if A an B could be diagonalized simultaneously, but this is impossible as stated earlier. In practice it is only possible to approximate the non-reflective boundary conditions.

For a better understanding of the derivation of the boundary conditions for the two-dimensional system we briefly look at the one-dimensional equivalent of system (7)

$$\vec{v}_t + \bar{A}\,\vec{v}_x + \bar{C}\,\vec{v} = \vec{0}, \tag{9}$$

where $\bar{A}$, and $\bar{C}$ are derived from $\tilde{A}$, and $\tilde{C}$ by dropping the second row and column. In contrast to the two-dimensional system, the left and right going waves that constitute the solutions of system (9) are uncoupled if the bottom friction is neglected, i.e. the components of $\vec{v}$ are described by uncoupled equations. If the domain of interest is $x \in [0,L]$, then the first component of $\vec{v}$. is incoming at $x = 0$ and outgoing at $x = L$ and visa versa for the second component of $\vec{v}$. A non-reflective inflow boundary condition at $x = 0$ would be

$$u + \psi = f(t). \tag{10}$$

Equation (10) prescribes the so-called ingoing Riemann-invariant.
If bottom friction is taken into account them the components of $\vec{v}$ are coupled. If, however, λ is small compared to the circular frequency, ω, of the waves under study, it is possible to make a series expansion of the solution in terms of the small parameter λ/ω. Effectively the same result is obtained by introducing new variables which, to a certain order of approximation, have uncoupled solutions. For the derivation of these approximations we take the Fourier transform of the system of equations (9) with respect to time and frequency variable ω. The method of frozen coefficients is used in order to get constant matrices in (9). Then (9) becomes

$$\hat{v}_x = - \bar{A}_f^{-1} \; (i\omega I + \bar{C}_f) \; \hat{v} = G \; \hat{v}, \tag{11}$$

where $\hat{v}$ is the Fourier transform of $\vec{v}$ and $\bar{A}_f$ and $\bar{C}_f$ refer to the frozen coefficient matrices $\bar{A}$ and $\bar{C}$.
G can be diagonalized with the aid of the matrix P formed by the eigenvectors of G as column vectors

$$P \; \hat{v} = (\; P \; G \; P^{-1}) \; P \; \hat{v}. \tag{12}$$

Since PGP^{-1} is a diagonal matrix, the components of $P \hat{v}$ are uncoupled. The boundary conditions are non-reflective when the quantities $P \hat{v}$ are prescribed at ingoing characteristics. Components of $P \hat{v}$ related to the negative diagonal values $P \; G \; P^{-1}$ are prescribed at $x = 0$ and those related to the positive diagonal values at $x = L$. In order to find the components of $(P \hat{v})$ in physical space one has to make a back transformation. This can in general only be done upto some order of approximation in the parameter λ/ω. To that end P is expanded into a Taylor series.
The zeroth-, and second-order outflow conditions at $x = L$ derived in [3,4] by taking one or three terms in the Taylor expansion, are given by

0 - th order

$$u - \psi = f(t), \tag{13a}$$

2 - nd order

$$(u - \psi)_t + \frac{\lambda}{4} \{2(u - \psi) + (1 + \frac{2u}{\psi}) (u + \psi)\} = g(t). \tag{13b}$$

The derivation of the weakly-reflective boundary conditions for the two-dimensional case can be carried out along the same lines, i.e.

- freeze the coefficient matrices,
- make a Fourier transform in y and t and write the system in a form similar to equation (12),
- determine the eigenvectors of the right-hand-side matrix and the transformation matrix P,
- make an approximate back-transformation to physical space.

Unfortunately, the eigenvectors cannot be obtained in explicit form for λ and f simultaneously different form zero. As pointed out by Majda and Engquist, [5], one can instead of approximating the back-transformation approximating the right-hand-side matrix of equation (12) such that a back transformation if feasible.
For system (7) the transformation matrix P is written as

$$P = \sum_{\substack{p=0 \\ q=0}}^{\infty} \left(\frac{\eta}{\omega}\right)^p \left(\frac{1}{i\omega}\right)^q P_{pq}, \tag{14}$$

where η is the dual variable of y.
Various approximations of weakly-reflective boundary conditions can be derived for various values of the exponents p and q. Three sub-critical outflow contidions for a boundary parallel to the y-axis at x = L read, [6,7].

p = 0 and q = 0

$$u - \psi \cong f_1(t), \tag{15a}$$

p = 1 and q = 0

$$(u - \psi)_t - uv_y = f_2(t), \tag{15b}$$

p = 1 and q = 1

$$(u - \psi)_t + \frac{\lambda}{4} \{2(u - \psi) + (1 + \frac{2u}{\psi})(u + \psi)\} - \frac{2u}{\psi} fv - uv_y = f_3(t). \tag{15c}$$

Conditions for x = 0 and for boundaries parallel to the x-axis are easily derived by rotation of the variables and are explicitely given by Mooiman, [8].

REMARKS ON STABILITY AND VISCOUS TERMS

For a stable computation, the differential equations together with the boundary conditions must constitute a well-posed problem. The zeroth-order conditions, i.e. prescribing the Riemann-invariants for the equivalent one-dimensional problem, are known to provide a well-posed problem, [2] and [9], but for the higher order approximations no prove exists. If, however, the boundary condition is written as

$$(u - \psi)_t + F(u - \psi, u + \psi) = f(t), \tag{16}$$

where F (,) is some function of its arguments, it is conjectured that this boundary condition is stable iff the ordinary differential equation- for u or ψ has nosolution that grows exponentially in time. It is easily verified that all conditions specified above satisfy this condition.
For the one-dimensional problem one can derive several first order relations that do not fulfill equation (16) and prove to be unstable in numerical computations [3].

In all equations used so far, viscous terms were neglected. If these terms are taken into account, system (7) becomes an incomplete parabolic system. The only hyperbolic nature left is due to the continuity equation of water. In principal the derivation of weakly-reflective boundary conditions can proceed along the same lines as before, but one has to make sure that the boundary conditions are such that no boundary layer is built up near open boundaries [2]. For short wave problems (Laplacian in the inner area and a hyperbolic free surface condition) it is known that most conditions derived are unstable [10].

NUMERICAL REFLECTIONS

For the one-dimensional problem all waves are truly independent, if the bottom friction is neglected. In this case all approximations to the boundary conditions are the same and equal to the zeroth-order conditions, i.e. the ingoing Riemann-invariant. When applied in a numerical solution the reflection will generally differ from zero. This is due to the fact that non-reflective boundary conditions for the differential equations are not necessarily non-reflective for the numerical scheme level. This can be demonstrated by considering the linearized one dimensional system

$$u_t + g\zeta_x = 0, \tag{17}$$

$$\zeta_t + hu_x = 0.$$

The (independent) Riemann-invariants for system (17) read

$$u \pm \sqrt{\frac{g}{h}}\,\zeta. \tag{18}$$

When the box or Preissmann-scheme, [11], is used to solve system (17) the finite difference approximations read

$$u_{i+1}^{n+1} + u_i^{n+1} + rg\,(\zeta_{i+1}^{n+1} - \zeta_i^{n+1}) = u_{i+1}^n + u_i^n - rg\,(\zeta_{i+1}^n - \zeta_i^n), \tag{19}$$

$$\zeta_{i+1}^{n+1} + \zeta_i^{n+1} + rh\,(u_{i+1}^{n+1} - u_i^{n+1}) = \zeta_{i+1}^n + \zeta_i^n - rh\,(u_{i+1}^n - u_i^n),$$

with $r = \Delta t/\Delta x$.
The general solution can be written as

$$\binom{u}{\zeta}_i^n = a_1 \binom{1}{\sqrt{\frac{h}{g}}} T^n X^i + a_2 \binom{1}{-\sqrt{\frac{h}{g}}} T^n X^{-i}, \tag{20}$$

were

$$X = \frac{\alpha-1}{\alpha+1}, \qquad \alpha = r\sqrt{gh},$$

and the time dependent function T as well as the constants a_1 and a_2 are determined by the boundary conditions.
If the discretized boundary conditions are applied one can easily verify that a_1 is only determined by the boundary condition at $x = 0$ and a_2 by the condition at $x = L$. The discretized formulation has conserved the non-reflectiveness of the continuous formulation. Numerical experiments with the Preissmann-scheme suggest that this property is conserved also for the full quasi-linear one dimensional equations, [4].

Following the same type of analysis it can be shown that space staggered schemes can not preserve the non-reflective property. A Hansen-type of discretization, Fig. 1, can be written as

$$u_i^{n+1} = u_i^n - rg\,(\zeta_{i+\frac{1}{2}}^n - \zeta_{i-\frac{1}{2}}^n), \tag{21}$$

$$\zeta_i^{n+1} = \zeta_i^n - rh\,(u_{i+\frac{1}{2}}^{n+1} - u_{i-\frac{1}{2}}^{n+1}).$$

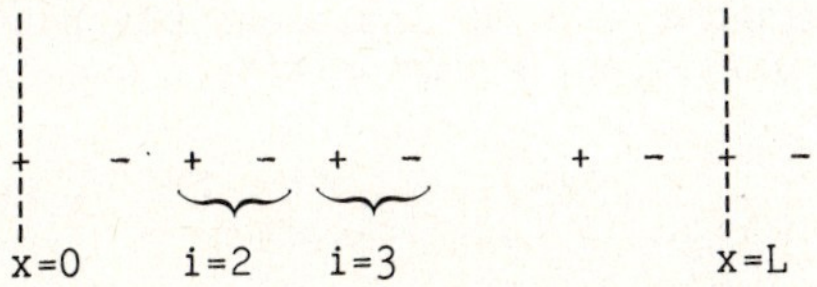

x=0 i=2 i=3 x=L

Figure 1. Grid configuration space staggered scheme.

The simplist form of the boundary condition at x=0 is given by

$$u_1^n + \sqrt{\frac{g}{h}}\,\zeta_1^n = f(t).$$

An analogue analysis of the Preissmann-scheme shows that a_1 and a_2 depend on the time dependent boundary condition at both sides, i.e. the left and right going waves which are uncoupled in the continuous formulation are coupled in the discrete formulation. It can be shown that the property of non-reflectiveness can only be preserved if the discrete formulation has the same eigenvectors as the differential equations.

A FINITE ELEMENT IMPLEMENTATION

The implementation of weakly-reflective boundary conditions is both possible for implicit as explicit finite element programs. However, the implementation for explicit methods is far more easy, since explicitness implies that boundary conditions may be incorporated after the time step has been carried out. The zeroth- and first-order approximations, equations (15a) and (15b), of the non-reflective boundary conditions, are implemented in the finite element package AFEP [7]. For a fast solution it is necessary to write system (1) in conservative form (i.e. with independent variables $u\ (h+\zeta)$, $v\ (h+\zeta)$ and $(h+\zeta)$ instead of u, v and ζ), and to use the so-called product approximation [13].
The product approximation is a variant of the standard Galerkin approach allowing for time-independent matrices, and thus an enormous reduction in time is obtained for building matrices and vectors. For the space discretization isoparametric quadrilaterals are used, with bilinear basis functions on the reference element. The time integration is due to Sielecki [13]. The mass matrix is lumped in order to get a fully explicit time scheme characterised by:

$$\begin{aligned} U^{n+1} &= f_1\ (U^n,\ V^n,\ H^n), \\ V^{n+1} &= f_2\ (U^{n+1},\ V^n,\ H^n), \\ H^{n+1} &= f_3\ (U^{n+1},\ V^{n+1},\ H^n), \end{aligned} \tag{22}$$

with $U = u(h+\zeta)$, $V = v(h+\zeta)$, and $H = (h+\zeta)$.

There are several possible ways to implement the boundary conditions. For the zeroth-order approximation, equation (15a), one can use for example

$$u^{n+1} = \bar{\psi}^{\,n+1} + f(t) \qquad \text{or} \qquad \psi^{n+1} = \bar{u}^{n+1} - f(t),$$

where $\bar{u}^{n+1}$ and $\bar{\psi}$ are the results of equation (22) without effect of boundary conditions.
Either of these formulations can be combined with the condition

$$u^{n+1} - \psi^{n+1} = \bar{u}^{n+1} - \bar{\psi}^{n+1}, \tag{23}$$

which guarantees that the equivalent one-dimensional Riemann-invariant is conserved at $t = (n+1)\ \Delta t$.
An other set of implementations is possible by taking into account that in the Sielecki-scheme U and V lag Δt in time relative to H. It is not possible to deside which implementation will give the smallest reflections without making a (rather complicated) numerical analysis or making the actual computation.
Slob, [7], investigated all combinations mentioned above and found that the smallest reflection coefficient is obtained if the time lag of U and V in the Sielecki-scheme is accounted for.
The example used to investigate the effectiveness of the weakly-reflective boundary conditions and their numerical implementation is the evolution of an initially Gaussian-shaped free surface elevation, Fig. 2A.

The relevant parameters are : $h = 10m$, $\lambda = 0$, $f = 0$, and

$$\zeta\ (x,\ y,\ 0) = \exp\ (x^2 + y^2)/L^2,$$

with $L = 200m$. The numerical parameters are $\Delta x = 50m$ and $\Delta t = 5s$, so the Courant number is just below one.
To study the reflections in more detail the solution in a limited domain is subtracted form the solution in a much larger domain and a reflection coefficient, Re, is defined as

$$Re = \frac{\max_t\ (\Delta\zeta)}{\max_t\ (\zeta)}\ . \tag{24}$$

Though Re is not a proper reflection coefficient in that its value is bounded to the range [-1,1], it provides a measure for the reflections.

In Fig. 2 the evolution of the free surface elevation is given at four instances of time. It is important to observe that the numerical solution remains circular symmetric and no indication of distortion in any direction is visible.

In Fig. 3 the reflection coefficient, Re, is given as a function of the angle of incidence of the wave impinging on the open boundary (in this case proportional to the coordinate along an open boundary).

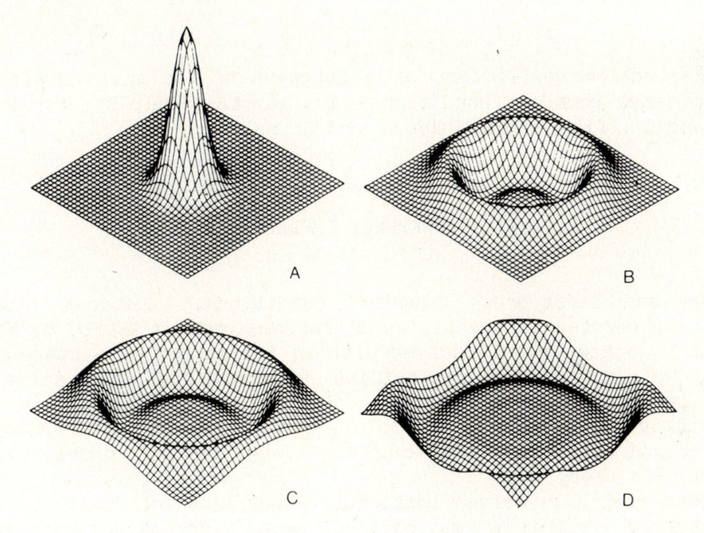

Figure 2. Time evolution of an initial, Gaussian-shaped, free surface elevation at t = 0, 60, 80 and 110 s.

The reflection coefficient depends rather strongly on the angle of incidence and increases about a factor of three in the range (0° - 45°). At normal incidence the reflecton coefficient is about 7%, this is the combined result of numerical reflections generated by the numerical scheme

itself and the approximation of the boundary conditions.
As the results of the zeroth and first-order approximations are about the same and inview of earlier (one-dimensional) experiences one might expect that a large fraction of the reflection at normal incidence is due to numerical reflections. These reflections must be decreased substancially before a higher than zeroth-order condition is useful.

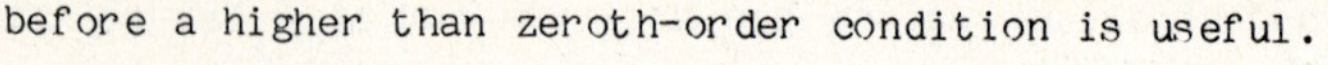

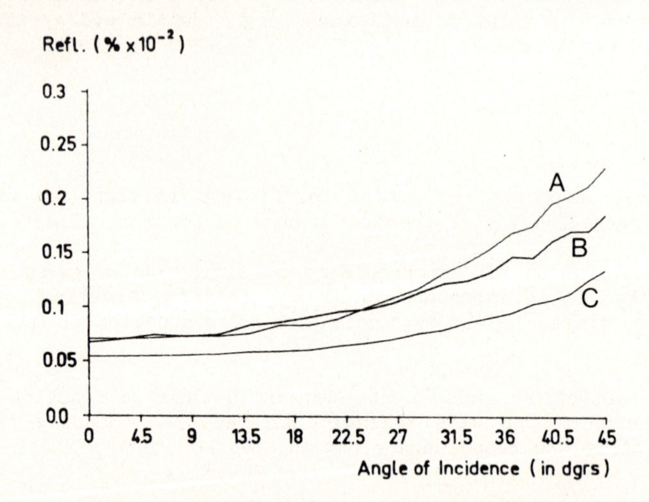

Figure 3. Reflection coefficient as a function of the angle of incidence and the boundary conditions. A : zeroth-order, B : first-order, and C : first-order with Δx and Δt halved.

A FINITE DIFFERENCE IMPLEMENTATION

The zeroth- and first-order boundary conditions, equations (15a) and (15b), are implemented in the finite difference program DELFLO by Mooiman, [8]. In DELFLO system (1) is solved with an ADI-method on a space staggered grid. For this paper it suffice to know that the solution at $t=(n+1)\ \Delta t$ is obtained in two steps:
in the first step v is solved implicitly along lines of constant x, and u and ζ are solved implicitly (and coupled) along lines of constant y. For v the new value is used.
in the second step u is solved implicitly along lines of constant y, and v and ζ are solved implicitly (and coupled) along lines of constant x. For u and ζ the new value is used.
The solution obtained after these two steps is second order accurate in space and time [12]. Formally, only the free surface elevation or the flow components could be prescribed at open boundaries. At inflow boundaries the tangential velocity component and the advective terms of the momentum equations in the direction normal to the open boundary were set to zero. At outflow the tangential velocity component was obtained by extrapolation and the advective terms normal to the open boundary was approximated by one-sided differences.

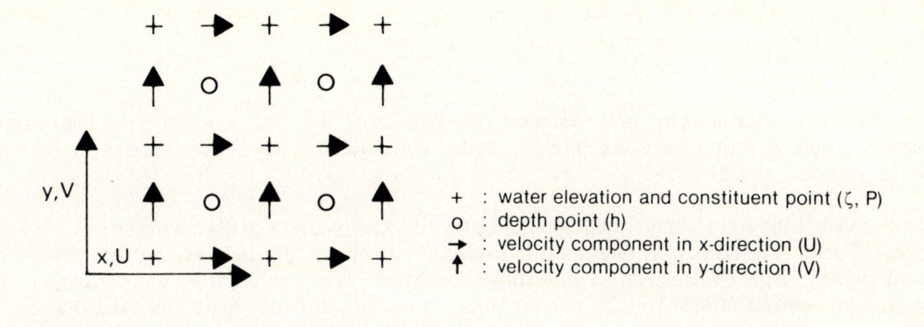

Figure 4. Space staggered grid of DELFLO.

The weakly-reflective boundary conditions are applied at the u or v-point. In the direction normal to an open boundary the momentum equation is used to arrive at the required number of equations.
At an inflow boundary normal to the x-axis for the zeroth-order condition results in

- first step

$$u_{i,j}^{n+1/2} + 2\sqrt{g(\bar{h}_{i,j}^{y} + \bar{\zeta}_{i.j}^{x^{n+1/2}})} = f_j^{n+1/2}, \tag{25a}$$

$u_{i,j}^{n+1/2}$ is found from the u-momentum equation at the boundary point with u_x set to zero, and solved implicitly with equation (25a),

$v_{i,j}^{n+1/2}$ is found from the second inflow condition

$$v_{i,j}^{n+1/2} = g_j^{n+1/2}, \tag{25b}$$

- second step

no boundary condition is required to find $u_{i,j}^{n+1}$, because ζ_x is evaluated, as before, at the old time level,

$v_{i,j}^{n+1}$ is found by applying the boundary condition

$$v_{i,j}^{n+1} = g_j^{n+1}, \tag{25c}$$

and $\zeta_{i,j}^{n+1}$ is found from the zeroth-order condition

$$u_{i,j}^{n+1} + 2\sqrt{g(\bar{h}_{i,j}^{y} + \bar{\zeta}_{i,j}^{x\,n+1})} = f_j^{n+1} . \qquad (25d)$$

For outflow boundaries and boundaries parallel to the x-axis the implementation goes along the same lines. This concludes the implementation of the zeroth-order condition.

More complicated is the implementation of the first-order condition, equation (15b). In fact, there exist several ways to arrive at an in time second-order approximation for inner region and boundary conditions. In order to avoid additional iterations, the equations must be solved in a very specific sequence. In the method used the boundary points are updated first, followed by the inner points. This procedure is used in both steps; full details are given by Mooiman [8].

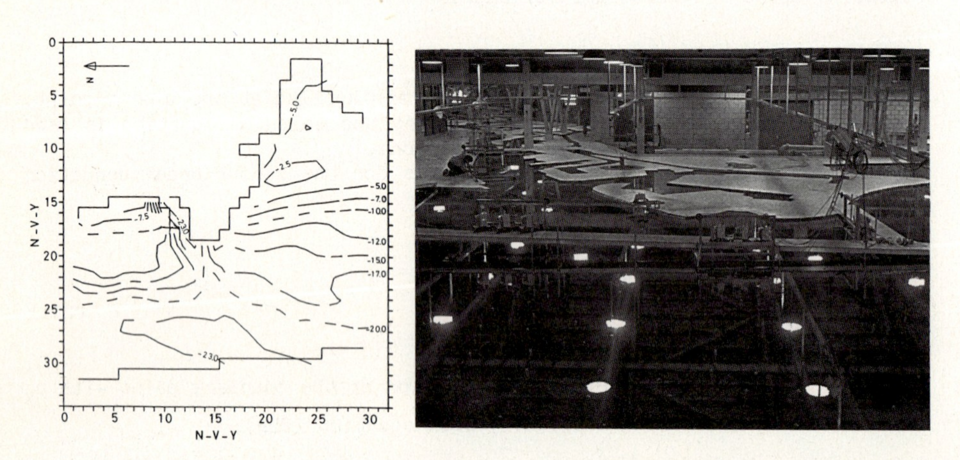

Figure 5. Hydraulic scale and mathematical model of Rijnmond.

The first example is the time evolution of the same free surface elevation as used in the finite element code. For the same set of parameters the result are about the same and not repeated here.
The second example concerns the influence of weakly-reflective boundary conditions at the transient. As explained in the introduction the transient can be quite large if initial disturbances only decrease by bottom friction. With weakly-reflective boundary conditions one might expect a much shorter transient, which in the limit of non-reflective boundaries decrease to the time it takes a wave to cross the domain.

The region modelled in this example is a part of the North Sea in front of Hook of Holland. Of this region both a hydraulic scale and a mathematical model exists, Fig. 5. The western boundary is closed as it coincides with a streamline. In the mathematical model generally water levels are prescribed at the southern and northern open boundary. The flow exchange with the Rotterdam Waterway is neglected in these computations.

In Fig. 6A the results are given for the water level at one station in the centre of the domain. At the open boundaries the water level (M_2 compo-

nent) is prescribed. The transient is characterized by strong eigen-oscillations which are damped in about one tidal cycle, i.e. 12 hours.
In Figs. 6B and 6C results are given for the zeroth- and first-order boun

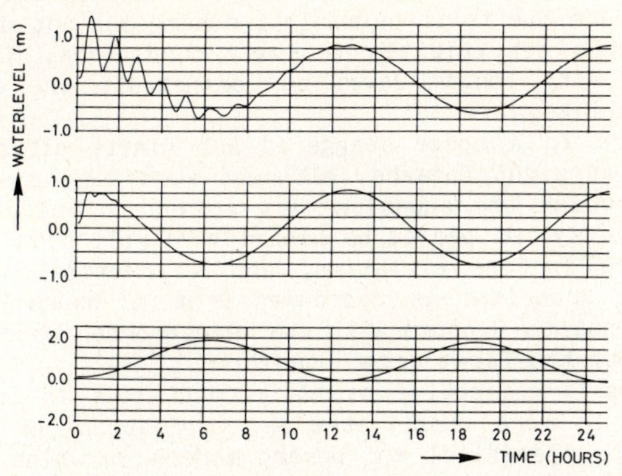

Figure 6. Water level at station (22,19) for different boundary conditions:
A: water level; B: zeroth-order; C: first-order.

dary conditions. The transient decreases to roughly 3 hours for the zeroth-order condition and to zero for the first-order condition. These results demonstrate the effectiveness of weakly-reflective boundary conditions in getting rid of (initial) disturbances.
The amplitude and phase of the resulting free surface elevation sligthly differ for 6A-6C. This is because the right-hand side of the boundary conditions are not fully adjusted to give the same results. For this, both velocity components and the free surface elevation must be known at the boundary. If only the free surface elevation is available, as in this example, a few iterative computations are necessary to adjust the right-hand side of the boundary conditions. This was not required here.

DISCUSSION

The effectiveness of weakly-reflective boundary conditions depends strongly on the small parameters λ/ω and f/ω, i.e. bottom friction and Coriolis coefficient over circular frequency, and η/ω, which is a measure for the angle of incidence. In the numerical solution the Courant number is important and is found in the reflection coefficient derived for simplified equations [9]. An excellent test case for one-dimensional equations can be found by neglecting the bottom friction. The left and right going waves are then uncoupled and reflections found in het numerical solution are due to the numerical scheme only. It is shown that the box or Preissmann scheme preserves this non-reflectiveness, but space staggered grids do not. For two-dimensional problems such a simple example does not exist as even in the case of vanishing bottom friction the waves remain coupled and reflections found in the numerical solution are from the numerical scheme and from the approximations of the boundary conditions.
The implementation in finite element codes is found to be straight for-

ward. For the time evolution of a Gaussian free surface elevation, which basically consists of relatively short waves, reflection coefficients of five to fifteen percent are found. It is expected that most of these resudial reflections are due to the numerical scheme and not due to the approximations made when deriving the boundary conditions. More computations need to be done with longer waves to demonstrate the efectiveness for practical applications.
The implementation in a space staggered ADI finite difference code also was found to be straight forward, althougt some care must be taken with the sequence in which the equations are solved to maintain second-order time accuracy. The system proved to behave very well both for short waves, the Gaussian free surface elevation, and a practical application with tidal waves. The transient is decreased from 12 hours, with reflective boundary conditions, to 3 hours with the zeroth-order condition and effectively to zero with the first-order condition.
A drawback of weakly-reflective boundary conditions is that the boundary forcing function is more complicated as combinations of water level and velocities must be specified. For nested models in which the results are taken from a larger domain model this of course is no problem. If the boundary conditions are taken from measurements, or if only the free surface elevation is known a few iterative computations are necessary to adjust the boundary forcing function.
A final remark concerns the practical use to this type of boundary conditions. From a physical point of view they can only be used if the outer domain behaves indeed as a perfect absorber. If not, one must include the reflections from the outer domain into the boundary forcing function by specifying a relation between the in- and outgoing waves. For the zeroth-order conditions this would simply be a combination of the Riemann-invariants. In some situations it is even not allowed to use weakly-reflective conditions, for instance when nature itself strongly reflects outgoing waves. This occurs when there is an abrupt change in the geometry, bottom friction, bottom topography, etc. Either, one chooses the boundary outside this domain or one prescribes the water level or flow velocities (and might have serious problems in calibrating the model).

Acknowledgment
The autors are indepted to J. Pakvis, A. Slob, and J. Mooiman for allowing them to use their results in this paper. These results were obtained as partly fulfilment of obtaining their Master Degree in Applied Mathematics from the Delft University of Technology during a nine month term of probation at Delft Hydraulics.

REFERENCES

[1] Daubert, A. and Graffe, O.: Quelques aspects des écoulements presques horizonteaux a deux dimensions en plan et non-permanents application aux estuaries. La Houille Blanche (1967), 8, p.p. 847-860.
[2] Oliger, J. and Sündström, A.: Theoretical and practical aspects of some initial boundary value problems in fluid dynamics. SIAM (1978), 35, p.p 419-446.
[3] Pakvis, J.: Weakly-reflective boundary conditions for shallow water equations (in Dutch), (1983). Ms. Sc. Thesis Delft University of Technology and Delft Hydraulics, Report S 545-1.
[4] Verboom, G.K.: Weakly-reflective boundary conditions for the shallow

water equations. Delft Hydraulics, (1982), Publication no. 266.

[5] Majda, A. and Engquist, B.: Radiation boundary conditions for acoustic and elastic wave calculations, Comm. Pure Appl. Math. (1979), 32, p.p. 313-357.

[6] Verboom G.K. and Slob, A.: Weakly-reflective boundary conditions for two-dimensional shallow water problems. Proc. 5th Int. Conf. Finite Elements in Water Recources, (1984), Burlington, USA.

[7] Slob, A.: Weakly-reflective boundary conditions for shallow water equations in two dimensions, (1983). Ms. Sc. Thesis Delft University of Technology and Delft Hydraulics, Report S 545-2.

[8] Mooiman, J.: Weakly-reflective boundary conditions and numerical implementation in DELFLO (in Dutch), (1986). Ms. Sc. Thesis Delft University of Technology and Delft Hydraulics, Report S 545-3.

[9] Verboom, G.K., Stelling, G.S., and Officier, M.J.: Boundary conditions for the shallow water equations. In Engineering Applications of Computational Hydraulics: Homage to Alexandre Preissmann (Edited by Abbott, M.B; and Cunge, J.A;). Pitman, London, (1982).

[10] Romate, J.: Private communications, (1985).

[11] Cunge, J.A. and Liggett, J.A.: Numerical Methods of Solution of the Unsteady Flow Equations. In Unsteady Flow in Open Channels, Volume 1, edited by Mahmood, K. and Yevjevich, V., (1975). Water Recources Publ., Fort Collins.

[12] Stelling, G.S.: On the construction of Computational Methods for shallow water flow problems, (1983), Ph. D. Thesis. Delft University of Technology, Dept. of Mathematics.

[13] Segal A., Praagman N.: A fast implementation of explicit time-stepping algorithms with the finite element method for a class of nonlinear evolation problems, Int. J. for Num. Meth. in Engineering, 23, pp. 155-168 (1986).

Addresses of the editors of the series „Notes on Numerical Fluid Mechanics":

Prof. Dr. *Ernst Heinrich Hirschel*
Herzog-Heinrich-Weg 6
D-8011 Zorneding
FRG

Prof. Dr. *Keith William Morton*
Oxford University
Numerical Analysis Group
8–11 Keble Road
Oxford OX1 3QD
Great Britain

Prof. Dr. *Earll M. Murman*
Department of Aeronautics and Astronautics
Massachusetts Institut of Technology (M.I.T.)
Cambridge, MA 02139
U.S.A.

Prof. Dr. *Maurizio Pandolfi*
Dipartimento di Ingegneria Aeronautica e Spaziale
Polytechnico di Torino
Corsa Duca Degli Abruzzi, 24
I-10129 Torino
Italy

Prof. Dr. *Arthur Rizzi*
FFA Stockholm
Box 11021
S-16111 Bromma 11
Sweden

Dr. *Bernard Roux*
Institut de Mécanique des Fluides
Laboratoire Associé au C.R.N.S. LA 03
1, Rue Honnorat
F-13003 Marseille
France